43 iwg 120
lbf 45402

Ausgeschieden im Jahr 2025

TRANSISTORS
FROM CRYSTALS TO INTEGRATED CIRCUITS

TRANSISTORS
FROM CRYSTALS TO INTEGRATED CIRCUITS

M Levinshtein & G Simin
IOFFE Institute of Russian Academy of Sciences, Russia

Translated by
Minna M Perelman

World Scientific
Singapore • New Jersey • London • Hong Kong

Published by

World Scientific Publishing Co. Pte. Ltd.
P O Box 128, Farrer Road, Singapore 912805
USA office: Suite 1B, 1060 Main Street, River Edge, NJ 07661
UK office: 57 Shelton Street, Covent Garden, London WC2H 9HE

British Library Cataloguing-in-Publication Data
A catalogue record for this book is available from the British Library.

TRANSISTORS. FROM CRYSTALS TO INTEGRATED CIRCUITS

Copyright © 1998 by World Scientific Publishing Co. Pte. Ltd.

All rights reserved. This book, or parts thereof, may not be reproduced in any form or by any means, electronic or mechanical, including photocopying, recording or any information storage and retrieval system now known or to be invented, without written permission from the Publisher.

For photocopying of material in this volume, please pay a copying fee through the Copyright Clearance Center, Inc., 222 Rosewood Drive, Danvers, MA 01923, USA. In this case permission to photocopy is not required from the publisher.

ISBN 981-02-2743-4

This book is printed on acid-free paper.

Printed in Singapore by Uto-Print

To my friends. M.L.
To my sons Kirill and Misha. G.S.

Preface

This book will tell you about the design and work of transistors and diodes — the most important semiconductor devices.

A few years ago this book was published in Russian in Moscow and about 140 000 copies of it were sold in the former USSR.

We found that the majority of those who had bought it were senior grade pupils interested in semiconductors, those who thought about their future careers, those who were considering semiconductor physics, materials science, electrical engineering or semiconductor processing as their future profession, though they didn't quite realize what semiconductor electronics actually is.

Many of our readers were undergraduates (mainly freshmen). We were glad to know that the book had proved to be useful supplementary reading, facilitating their learning process.

We did not expect it and were happy to know that graduates and postgraduates were also interested in the book. Some of our readers have degrees (B.Sc., or M.Sc., some are getting PhDs), and yet they wrote to tell us that what had seemed to them most complicated and perceivable only by means of differential equations and computer simulation, became quite simple and understandable after they had read the book. Semiconductor designers and technologists were also interested in the book. They found some new ideas and unexpected approaches in the book.

Several professors, who deliver lectures in semiconductor physics and semiconductor devices, have written to us saying that they think it is expedient to recommend the book to their students as supplementary reading.

Thus, the book has proved to be interesting to quite a lot of people whose range and levels of knowledge are very different.

We are thankful to our friends and colleagues: Michael I. Dyakonov, Professor, Principal Scientist of the Ioffe Institute, St. Petersburg, Russia; Boris I. Shklovskii, Professor, University of Minnesota, USA; Igor M. Filanovskii, Professor, University of Alberta, Canada for their help in the work featured in this book. We are grateful to our wives Larisa and Marianna for their patience, understanding and friendly support. We want to thank our children: Dina Levinshtein and Misha and Kirill Simin for their love which helped a lot. We are grateful to Minna M. Perelman, the translator of this book, and also of our book "Getting to Know Semiconductors", published by World Scientific in 1992. Her energy and courage inspired our work. Our special thanks are also extended to Marianna A. Simin: her funny pictures created an easy and friendly atmosphere.

M. Levinshtein
G. Simin

Contents

Preface vii

Introduction 1

Part I. Semiconductors 3

Chapter 1. The Main Properties of Semiconductors 7

1.1 Intrinsic Semiconductors 7
   1.1.1 "Free electrons" in crystal 8
   1.1.2 Holes 9
   1.1.3 Generation and recombination 10
      Thermal generation 11
      Electron-hole recombination 12
      Intrinsic concentration 13
      The simplest band diagrams 14

1.2 Impurity Semiconductors 16
   1.2.1 Donor impurity 17
   1.2.2 Acceptor impurity 19
   1.2.3 The temperature dependence of the carrier concentration 21
   1.2.4 Minority carriers 23
   1.2.5 Band diagrams 24
      Compensation 26

1.3 Deep Levels 28
   1.3.1 Compensation by deep levels 29
   1.3.2 Generation through deep levels 31
   1.3.3 Recombination through deep levels 34

1.4 Summary 36

Chapter 2. Motion of Electrons and Holes inside the Crystal 39

2.1 Thermal Motion 40
   2.1.1 Energy distribution of electrons 40
   2.1.2 Energy distribution of holes 42

2.2 Motion in the Electric Field	48
2.2.1 "Hot" electrons	50
2.2.2 Band diagrams	51
2.3 Diffusion	55
2.3.1 Diffusion coefficient	57
2.3.2 Diffusion current	59
2.3.3 Diffusion length	60
2.4 Summary	64
Part II. Barriers & Junctions	**65**
Chapter 3. The Barrier on the Crystal Boundary	**67**
3.1 Work Function	69
3.1.1 Come back, return, I call you back!	70
3.1.2 Double charged (dipole) layer	70
3.1.3 How to define the work function	72
Red boundary of the extrinsic photoelectric effect	72
Thermoionic emission	73
3.1.4 What work function is equal to	74
Work function in metals	74
Work function in semiconductors	74
3.2 Surface States	78
3.3 Bending Bands, Surface Potential	80
3.4 Summary	84
Chapter 4. The Main Parameters of Potential Barriers	**85**
4.1 How the Electric Field Penetrates into a Metal, Dielectric and Semiconductor	86
4.1.1 Why the electric field practically does not penetrate into a metal	86
4.1.2 How the electric field penetrates into a dielectric	88
4.1.3 In what way and how deep the electric field penetrates into a semiconductor	90
4.2 Field Dependence on the Coordinate	92
4.3 Poisson's Equation	94
4.4 A Few Words about Accumulation Layers	97
4.5 Summary	98

Chapter 5. *p-n* Junction — 99

5.1 Ways of Obtaining *p-n* Junctions — 99
   5.1.1 Alloying — 101
   5.1.2 Diffusion — 103
   5.1.3 Ion implantation — 105

5.2 Barrier on the Boundary — 108
   5.2.1 The height of the barrier — 111
   5.2.2 Depletion layer. Width of the barrier — 113
   5.2.3 Wonderful equilibrium — 117
   5.2.4 The reverse bias — 124
      Height and shape of the barrier — 126
      Reverse current — 127
      Barrier capacity — 130
   5.2.5 The forward bias — 133
      The height of the barrier — 133
      Forward current — 133
      Injection — 137

5.3 Summary — 143

Chapter 6. Diodes with the *p-n* Junctions — 145

6.1 Photodiodes — 145

6.2 Variable Capacitors — 149

6.3 Light-Emitted Diodes — 151

6.4 Solar Cells — 154

6.5 Rectifier Diodes — 159

6.6 Summary — 165

Part III. Transistors — 167

Chapter 7. Bipolar Transistors — 169

7.1 Principle of Operation of a Bipolar Transistor — 170
   7.1.1 Current amplification — 172
   7.1.2 Parable about what is main and what is minute — 175
   7.1.3 Speed of response of the transistor — 176

7.2 Some Words about the Types and Manufacturing of Bipolar Transistors — 185

7.3 The Simplest Transistor Circuits	191
7.4 Summary	197
Chapter 8. Field Effect Transistors	**199**
8.1 The Beginning	199
8.1.1 The main idea	199
8.1.2 Simple estimations	201
8.1.3 Old acquaintances	202
8.2 Maturity and Flourishing	203
8.2.1 JFET (p-n-junction field effect transistor)	203
8.2.2 Fortune favours the brave. MOSFET	205
8.3 Epitaxy	209
8.4 A Few Important Details	211
8.5 The Work of the FETs in Actual Regimes	213
8.5.1 The main parameters of FETs	217
Transconductance	217
Speed of response	220
8.6 FET as an Element of Electronic Circuits	221
8.7 Summary	223
Chapter 9. Transistors and Life	**225**
9.1 The First King	226
9.2 Ugly Duckling	228
9.3 Long Live the New King!	231
9.4 The King... Disappears. Long Live the New King!	233
9.5 Claimants to the Throne	239
Three-dimensional intergrated circuits	239
Semiconductor elements of the optical computers	240
Bioelectronics	240
Conclusion	241

Introduction

Where there is a will, there is a way.

It's a long way to Tipperary,
It's a long way to go,
It's a long way to little Mary,
To the sweetest girl I know...

Semiconductor devices are numerous, even the most important of them make dozens, with each of them, as a rule, having lots of species.

In the library you can see a great many reference books, hundreds of pages each. Those reference books mention only the main properties of the most widely used semiconductor devices. But they do not say anything of the physical principles of the work of those devices.

Those principles are given in other thick volumes, again hundreds of pages each. Finally, there are many very thick books describing the application of semiconductor devices. How is one to orientate himself among those numerous devices and principles?

The point is that the work of a great many devices is based on a rather limited number of ideas. Some of those are quite simple and evident, others require certain effort and patience. After you have perceived those ideas and principles, you can understand the design and principle of operation of practically any semiconductor device.

First, in Part I of this book, *Semiconductors*, we will get to know the main properties of semiconductors. It is most essential to know them in order to understand the work of any semiconductor device. Some devices are based on a remarkable property of semiconductors — their ability to change their electric resistance even at the slightest alteration of the environmental conditions. Therefore, measuring the resistance of semiconductors, one can measure the temperature and illumination, pressure and magnetic field, the electric field and the velocity of the motion of liquid or gas, the acceleration or the extent of smoke-screen in the premises. Such devices are widely used in households and industry, in scientific research and the military service. Millions and millions of such devices are produced every year.

The main properties of semiconductors and the work of the devices, based on those properties, have been discussed in detail in our book "Getting to Know Semiconductors".[1] However, even if you have read that book, we recommend you not to skip the first part of this book. This is because we widely use a simple and expressive technique here — the "language" of energy diagrams. That "language" was not employed when describing the most simple semiconductor devices, but it proved most useful when describing the work of diodes and transistors. The basis of that language is introduced in Part I of this book. However, in order to understand the principle of diodes and transistors, it is not enough just to know the properties of semiconductors. It is also quite essential to study very interesting, unusual and not very simple phenomena — the so-called *junctions*. It may be *homojunction*, a region inside the semiconductor, whose different parts are doped with different impurities; it may be *heterojunction*, a region emerging on the border between two different semiconductors. In the vicinity of those borders there appear regions with quite peculiar properties — energy barriers, which determine the work of diodes and transistors.

The properties of junctions will be discussed in Part II of this book. And here we will discuss various types of diodes.

The third part of this book is devoted to the principle of design, work and application of transistors.

---

[1] M. E. Levinshtein, G. S. Simin. Getting to Know Semiconductors. World Scientific. Singapore — New Jersey — London — Hong-Kong, 1992. ISBN 981-02-0760-3.

Part I

Semiconductors

> "The English physicist Cavendish has proved experimentally that water conducts electricity 400 million times worse than metals; nevertheless it is not a very bad conductor of electricity. Bodies which take the intermediate position between conductors and nonconductors are usually called SEMICONDUCTORS"
> Ivan Dvigubsky.
>
> *"Fundamentals of Experimental Physics"*, *1826*

There is such a striking variation in the conductivity of different substances that one can't help but be surprised at it. The conductivity of a good conductor, like silver, is about 10^{22} times greater than the conductivity of a good insulator, like glass. To better understand the difference between the conductivity of good metals and that of good insulators, one should realize that it is the same as the difference between the diameter of our galaxy and the length of 1 cm!

Though the atoms of any substance are electrically neutral, any atom, silver, copper, nitrogen, oxygen, germanium or silicon contains the same number of positively charged protons and negatively charged electrons. So, on the whole, every atom is neutral. The volume, filled with neutral particles, is incapable of conducting the electric current. Consequently, the volume filled with separate isolated atoms of any substance — silver, silicon or diamond — is in fact an ideal insulator.

But why do those substances, when they are in the solid state, have quite different conductivities? Solids seem to be composed of the same atoms.

Moreover, the atoms of one and the same substance, e.g. of carbon (C), depending on the type of crystal they form, can be either a very good conductor (graphite), or a perfect insulator (diamond).

The last example suggests that whether the solid is a metal or a dielectric depends not so much on the properties of the atoms forming the crystal, as on the types of the bonds of atoms in the crystal lattice of the solid body.

Figures 1 and 2 demonstrate two main types of crystal lattice. Figure 1 illustrates the main properties of metallic crystals. The crystal lattice is formed by positively charged ions, not by neutral atoms. While forming the lattice, each atom loses one valence electron, and these electrons (shown as black circles in Fig. 1) do not belong then to any specific ion of the metal. These electrons are said to be collectivized by the crystal and can move freely under the action of the external electric field. There are $\sim 10^{22}$ free carriers — electrons — in every cubic centimetre of the metal. It is no wonder that metals are perfect conductors.

The scheme of the crystal lattice of silicon, shown in Fig. 2, demonstrates another type of bond in a crystal. Every atom is linked with the neighbouring atoms by means of strong electron bonds.

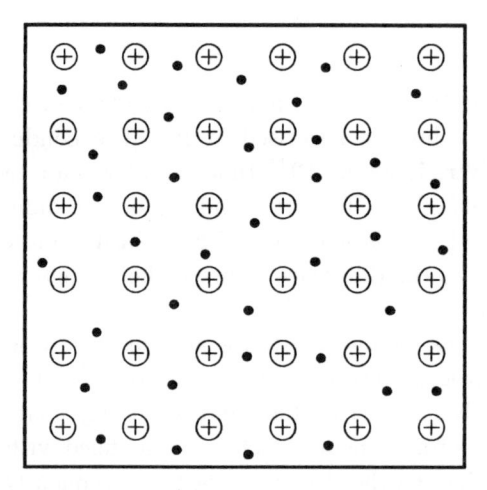

Fig. 1. Schematic diagram of the metal crystal lattice. The regular lattice of positively charged ions is plunged into the "gas" of free electrons, having no tight bonds with separate ions.

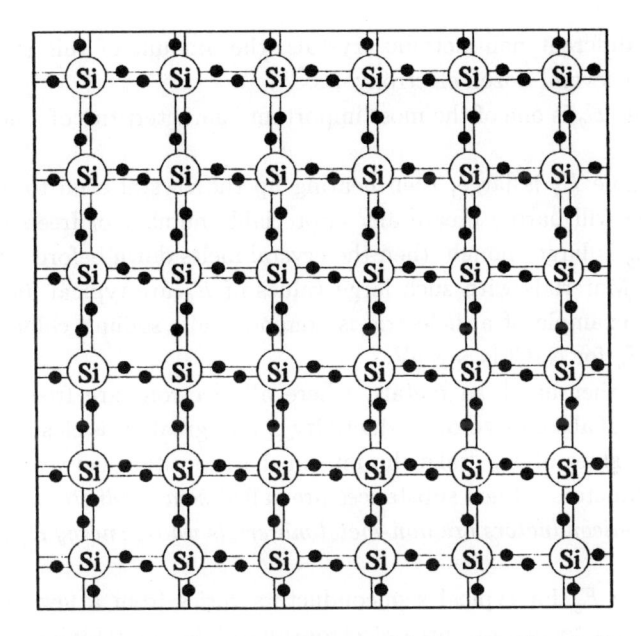

Fig. 2. Schematic diagram for the silicon (Si) crystal lattic. The lines linking the Si atoms represent electron bonds.

The field which keeps the electrons in their orbits is very strong. By order of magnitude they make $\sim 10^{10}$ V/m. Therefore it is not easy to break the electron coupling between the atoms and knock out the electron from the orbit. And while all the valent electrons are coupled on the interatomic orbits, there are no free electrons in the crystal i.e. there are no electrons able to move under the action of the external electric field within the crystal. Therefore, the ideal crystal, shown in Fig. 2, will be an ideal dielectric.

But alas, as we know from our own experience, ideal objects in the real world exist only in our imagination. Any defect of the crystal lattice, any foreign impurity, even heat, is able to break the electron bonds shown in Fig. 2 and release some electrons from the interatomic orbits. These released electrons can move under the action of the external electric field. So, in fact, every crystal can conduct electric current. How large that conductivity is depends on the number of released electrons. The latter depends on the fact that it is hard to break the links keeping the electron in the interatomic orbits.

The energy necessary to break the electron bond and free the electron is usually designated by the symbol E_g. The index "g" comes from the word "gap." In different non-metallic crystals, the amount of this energy varies from zero to several dozen electronvolts.*

The value E_g is one of the most important characteristics of a non-metallic crystal.

If the value E_g is large, then heating up the crystal even to a very high temperature will hardly create any appreciable number of free electrons. If the value E_g is large enough, then the crystal melts down before a free carrier is created. Materials with such large values of E_g are typical *dielectrics*. A well-known example of a dielectric is common salt, sodium chloride (NaCl). The value E_g for NaCl is ~ 8 eV.

On the other hand, in metals, where all electrons are free even at the temperature of absolute zero, conductivity is very great. It is clear that crystals with small values of E_g "take the intermediate position between conductors and nonconductors." Such substances are called *semiconductors*.

Thus, *semiconductors are non-metal materials whose energy E_g is relatively small*.

The value E_g for typical semiconductors varies from a few tenths of an electronvolt up to two or three electronvolts. Thus, for indium antimonide (InSb), $E_g = 0.17$ eV. For germanium (Ge), the material used to make the very first transistors and semiconductor diodes, $E_g = 0.72$ eV. For silicon (Si), which is the main material of the modern semiconductor electronics, $E_g = 1.1$ eV. Gallium arsenide (GaAs), the most promising material for semiconductor electronics of the near future, has $E_g = 1.4$ eV. In the ternary semiconductor compound GaAlAs used to make semiconductor light-emitting diodes and lasers, the value E_g varies from 1.4 eV (GaAs) to 2.17 eV (AlAs), depending on the relative amount of Al and Ga they contain. Silicon carbide (SiC), the material of the most reliable and stable light-emitting diodes, is able to operate at very high temperatures, has $E_g \approx 3$ eV.

We have named here only the most important semiconductor materials. But hundreds of semiconductor compounds have been synthesized, studied and made use of.

---

*A unit of energy of 1 electronvolt (1 eV) is equal to $1.6 \cdot 10^{-19}$ J. The electron acquires the energy of 1 eV after it passes the potential difference of 1 V. When we speak of the properties of solids, it's very convenient to measure the energy in electronvolts. We will frequently use this unit.

Chapter 1

The Main Properties of Semiconductors

1.1 Intrinsic Semiconductors

Intrinsic semiconductors are semiconductors in which the concentration of free current carriers is defined only by temperature and by the amount of energy E_g unique to the given semiconductor.

A logical question can be asked: what else can determine the concentration of free carriers? As we will see further, one of the most characteristic features of semiconductors is their very high sensitivity to a very very small quantity of impurity. A quite negligible (and inappreciable in any other materials) addition of a foreign substance — one atom per million or even milliard of a semiconductor can change its properties to such an extent that the concentration of the free carriers will be determined not by the intrinsic properties of the semiconductor, but by the amount and characteristic of the impurity which was added.

Let us imagine the most simple situation: the semiconductor crystal has no defects and does not contain any impurity. In the ideal crystal lattice, every electron is kept in its orbit, as shown in Fig. 2, and it will take the energy E_g to release it. The temperature of the crystal is T. What portion of the electron bonds will be broken? Or, how many electrons capable of moving under the action of the field, i.e. *conduction electrons*, will there be in the crystal?

Chaotic thermal motion tends to break the bonds between the atoms. The value of the energy of thermal motion is known to be kT, where k is the Boltsman constant. The value of $k = 1.38 \cdot 10^{-23}$ J/K or $8.6 \cdot 10^{-5}$ eV/K. At room temperature (300 K), the energy of thermal motion $kT \approx 0.026$ eV. When the temperature is rather high — +200°C or $\approx$ 500 K, the energy $kT = 0.043$ eV. But as we know, the value E_g for typical semiconductors is much higher. It may seem that in this case the thermal motion will be incapable of breaking a single electron bond and, provided $kT \ll E_g$, there will be no free carriers in the crystal.

However, the value kT characterizes only the average thermal energy of particles. Due to the chaotic character of thermal motion at any given moment of time, there will be such atoms in the crystal whose energy will prove greater than the value of E_g. Such atoms will not be numerous. But they will exist, and a part of the electron bonds between the atoms will be broken. What will take place is shown in Fig. 3. One of the links is broken (between atoms 19 and 20). The electron, being dislodged from its orbit, is within the area formed by atoms 1, 2, 7 and 8.

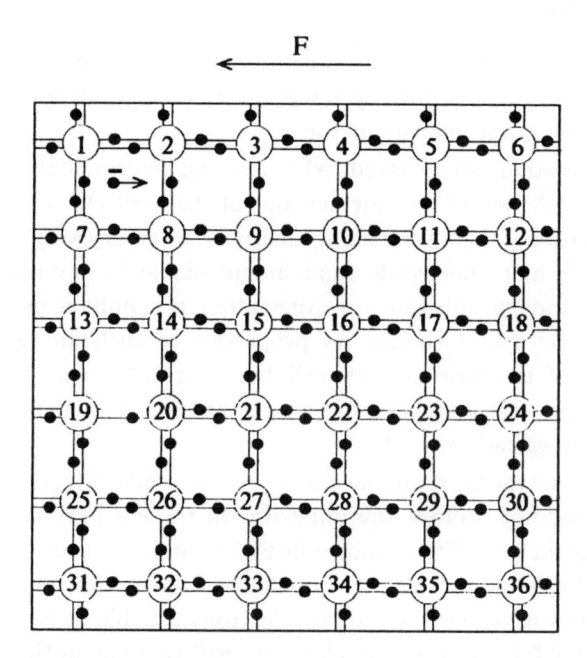

Fig. 3. One of the electron bonds (between atoms 19 and 20) is broken by chaotic thermal motion. A conduction electron appeared (it is located between atoms 1, 2, 7 and 8). The negatively charged electron will move against the electric field F (to the right).

1.1.1 "Free electrons" in crystal

From Fig. 3, it is clear that when speaking about an electron dislodged from its orbit, we should preferably use the word "free", at least mentally, in inverted commas.

The electron which is actually free is the one in the vacuum.

The so-called "free" electron in the crystal is in fact in the complex electric fields. The electric fields are formed by the ions of the lattice and by the valence electrons of the neighboring atoms.

Under the action of the external electric field, F, a free electron in the vacuum moves with a constant acceleration $a = qF/m_0$ (q is the electron charge, m_0 is the mass of the free electron). The "free" electron in a crystal can move freely only for a very short period of time, after which it is sure to collide with the atom of the lattice (Fig. 3). So, when using the term "the free electron in a crystal," we should bear in mind that it denotes in fact just an ability to perform a directed motion under the action of the external electric field, thus conducting an electric current. That is why free electrons in a crystal are called *conduction electrons*.

1.1.2 *Holes*

When the electron bond is broken (Fig. 3), the crystal acquires not one, but two possibilities to conduct the electric current. Let us compare the behaviour of the conduction electron with that of the broken bond (in Fig. 3, it is shown between atoms 19 and 20). If no external field is applied to the crystal, the electron will travel chaotically between the atoms of the lattice under the action of thermal oscillations. And what about the broken bond?

Any of the electrons linking atoms 19 and 20 with the adjacent atoms may get to the trajectory of the dislodged electron and thus restore the broken bond between atoms 19 and 20. If that happened to be the electron which had been linking, say, atoms 14 and 19, then the broken bond will take its position. Then it will be displaced to the position between atoms 9 and 14, or else between atoms 19 and 18. It may also move to the place between atoms 14 and 15. Just like the electron, the broken bond will travel chaotically between the atoms of the lattice.

If the crystal is acted upon by an external electric field, then apart from the chaotic motion, the free electron, as a negatively charged particle, will also acquire a directed motion against the electric field. What about the broken bond? Let us look again at Fig. 3. Now any of the electrons, linking atoms 19 and 20 with the adjacent atoms, can shift to the trajectory between these two atoms (i.e. atoms 19 and 20). However, due to the action of the external field, this shift is most probable for the electrons linking atoms 18 and 19. They would be "pulled" by the external field to the place of the broken bond. The directed motion along the field will be added to its chaotic motion. However, it does not mean that the broken bond will be necessarily substituted by the electron linking atoms 18 and 19 and that the hole will 1) sure to or 2) indispensably or 3) necessarily appear between atoms 17 and 18. The external electric field does not stop the chaotic motion of the broken bond. It just adds to the chaotic motion some elements of the directed motion. Note that the broken link moves in the direction opposite to that of electron's motion — it moves along the electric field. In other words, it behaves like a positively charged particle.

It goes without saying that the broken link is in fact not a true particle. Unlike the electron, it cannot be extracted from the crystal and studied in the vacuum. However, when discussing the properties of the crystal it is much easier to watch the displacement of a broken bond than the actual displacement of electrons from one orbit to another. A broken bond like this is called a *"hole"* and it is said to be a *quasiparticle*.

When a crystal is acted upon by an electric field, then a break in the link, a hole, is an additional source of conductivity. Electrons from the adjacent links jumping from one broken bond to another move against the field and conduct the current. If we are to define this mechanism of conductivity by means of the notion of a hole, we are to consider the hole, traveling in the field in the direction opposite to that of the electrons to be a *positively* charged particle whose charge is equal in magnitude to that of the electron.

Should we expect that the conduction electron and the hole will move with the same velocity in the electric field? Hardly so. Looking at Fig. 3, we can see that the conditions of the motion of a free electron within the crystal and the conditions of the motion of an electron from one bond to another are absolutely different. The motion from one orbit to another turns out to be much more difficult. In the same external field, the hole, as a rule, moves more slowly than the electron.

1.1.3 *Generation and recombination*

The chaotic thermal motion is steadily breaking electron bonds between the atoms and generates free electrons and holes. Is it possible that finally, as a result of that process, all the electron bonds will be broken and the semiconductor will become a metal? No, it is not!

Moving chaotically in the crystal, the free electron and the hole may happen to be close to each other. The free electron will take its place on the

Chapter 1. The Main Properties of Semiconductors   11

free trajectory of the interatomic link and ... both the free electron and hole disappear simultaneously. This process is called *recombination* which, when translated from Latin, means "reunion." In a steady state condition, there is a dynamic equilibrium: the number of electrons and holes (electron-hole pairs) generated every second in the volume of the semiconductor on account of thermal generation is equal to the number of those annihilated on account of recombination. A certain equilibrium concentration of electrons and holes is established in the semiconductor at a given temperature.

In crystals whose properties we study, electrons and holes are always created in pairs and they always perish in pairs. Therefore the concentration of electrons n_i is equal to the concentration of holes p_i (The index "i" comes from the word "intrinsic"). Let us determine the temperature dependence of the equilibrium concentration of the free carriers $n_i = p_i$.

To do it, we must determine how many free carriers are formed in one unit of volume of the semiconductor per unit of time and how many of them perish during the process of recombination. In the case of equilibrium, these two values must be equal to each other. Equating them to each other, we will determine the equilibrium concentration of the carriers.

Thermal generation. We have already learned the main feature of the electron-hole pair generation under the action of thermal motion. It takes much more energy than the average energy kT.

Investigating various physical processes we often come across situations where to make a certain event take place it is necessary to expend a large "portion" of energy ΔE, much greater than kT. That happens when we study the evaporation of water, or the decay of the nucleus; when we investigate the emission of the electrons by the cathode of the vacuum tube or the distribution of the density of gases in the atmosphere. In one of the main parts of physics, statistical physics, it is proven that the probability of such an event is always proportional to $\exp(-\Delta E/kT)$:

$$w \approx e^{-\frac{\Delta E}{kT}} \qquad (1)$$

where e is a number equal to 2.7183... It is called the natural logarithmic base.

The values of the energies $\Delta E = E_g$, necessary to break the electron bonds in various semiconductors, are known to us. Consequently, knowing the temperature T, we can calculate by Eq. (1) the probability of forming an electron-hole pair in different materials. At room temperature ($kT = 0.026$ eV), for InSb ($E_g = 0.17$ eV), this probability is proportional to

$e^{-6.54} \approx 1.44 \cdot 10^{-3}$. For Si ($E_g = 1.1$ eV), the probability is proportional to $e^{-42.3} \approx 4.23 \cdot 10^{-19}$; for GaAs ($E_g = 1.4$ eV), $e^{-53.8} \approx 4.12 \cdot 10^{-24}$.

Those examples illustrate the main feature of the exponential dependence: with the change of the index of the exponent, the value of the exponent changes a lot. The increase of ΔE by approximately 1.27 times diminishes the probability of the formation of an electron-hole pair by approximately 100 000 times! (cf. the probabilities for Si and GaAs). We will note that a similar change of the probability of forming electrons and holes will occur in one and the same semiconductor if the temperature is changed. So to decrease the probability of forming an electron-hole pair in Si by 100 000 times, it is enough to cool the crystal of silicon, lowering the temperature from room temperature to $-78°C$, i.e. to the temperature of dry ice, used to store ice-cream.

So, the number of electron-hole pairs K_1 being formed every second in a unit of volume of a semiconductor is

$$K_1 = \alpha e^{-\frac{E_g}{kT}} \qquad (2)$$

where α is a coefficient of proportionality, different for different semiconductors.

On the other hand, due to recombination, a certain number of charge carriers K_2 will disappear from the same unit of volume every second. So what does the number of recombining carriers depend on?

Electron-hole recombination. In order to recombine, the electron and hole must meet. What does the frequency of their meetings depend on? Let us mentally make the following experiment. We will watch some atom of the crystal lattice and mark the appearance of a hole in its vicinity. It is clear that the greater the number of holes per unit of volume of the semiconductor, the more frequently they appear there. The probability of the appearance of a hole is proportional to the concentration of holes p_i. For the same reason, the probability of the appearance of a free electron in the vicinity of that atom is proportional to the concentration of electrons n_i. We are interested in the probability of a simultaneous appearance of both the electron and the hole in the vicinity of the atom. This probability (i.e. the probability of their meeting and recombination) is proportional to the product of the concentrations of electrons and holes $n_i \cdot p_i$. Thus, the number of carriers recombining every second per unit of volume is the following:

$$K_2 = \beta \cdot n_i \cdot p_i. \qquad (3)$$

Chapter 1. The Main Properties of Semiconductors

The coefficient β as well as α in Eq. (2), is different for different semiconductors.

Since $n_i = p_i$, then

$$K_2 = \beta \cdot n_i^2 = \beta \cdot p_i^2. \tag{4}$$

Equating the number of the newly born pairs K_1 to the number of the perishing pairs K_2, we obtain:

$$\alpha e^{-\frac{E_g}{kT}} = \beta \cdot n_i^2 = \beta \cdot p_i^2. \tag{5}$$

Hence we obtain

$$n_i = p_i = \sqrt{\frac{\alpha}{\beta}} \cdot e^{-\frac{E_g}{2kT}} = (A \cdot B)^{1/2} \cdot e^{-\frac{E_g}{2kT}}. \tag{6}$$

Values A and B in Eq. (6) are measured in cm^{-3}. They are known for all the semiconductors we will deal with. At room temperature these values are within the limits approximately from 10^{17} to 10^{19} cm^{-3}. Knowing the values of A and B, it is easy to calculate by Eq. (6) the values of the concentration of intrinsic carriers in any semiconductor at any temperature.

Intrinsic concentration. Table 1 gives the values E_g and n_i for some semiconductors at room temperature (300 K).

Table 1

Semiconductor	InSb	Ge	Si	InP	GaAs	GaP	SiC
E_g, eV	0.17	0.72	1.1	1.3	1.4	2.3	2.4 – 3.2
Intrinsic concentration n_i, cm^{-3}	$1.3 \cdot 10^{16}$	$2.4 \cdot 10^{13}$	$1.1 \cdot 10^{10}$	$5.7 \cdot 10^7$	$1.4 \cdot 10^7$	0.8	$0.12 - 2.10^{-8}$

We remember that every cubic centimetre of metal contains $\sim 10^{22}$ conduction electrons. From Table 1, we can see that even in a semiconductor with a small value of E_g — InSb, the concentration of intrinsic electrons at room temperature is a million times smaller than that in a typical metal.

In Si, it is a trillion times (by a factor of 10^{12}) smaller. In GaAs, it is smaller by a factor of 10^{15}. And in silicon carbide (SiC), it is smaller by a factor of 10^{30}!

We can see from Eq. (6) that the concentration of the carriers, (and consequently the conductivity) grows abruptly with the rise of temperature. The values A and B, speaking generally, do not change much with the temperature. The main contribution into temperature dependence of the intrinsic concentration is made by an exponent. Table 2 gives the values of n_i at the temperature of 500 K for the same semiconductor materials:

Table 2

Semiconductor	InSb	Ge	Si	InP	GaAs	GaP	SiC
Intrinsic concentration n_i (cm^{-3}) at 500 K	$4.8 \cdot 10^{16}$	$6.4 \cdot 10^{15}$	$5.6 \cdot 10^{13}$	$1.4 \cdot 10^{12}$	$7.2 \cdot 10^{11}$	$4.4 \cdot 10^{7}$	$\sim 10^{2}$

We can see that the greater E_g is, the more abruptly the temperature of the intrinsic concentration grows. If the temperature increases 1.7-fold, the intrinsic concentration in InSb increases about 3.7-fold; in Si, about 5 000-fold; and in SiC, by a factor of 10^{10}!

An abrupt growth of conductivity with the rise of temperature is one of the most characteristic properties of intrinsic semiconductors.

The simplest band diagrams. There is a very simple and easy method, giving a qualitative description of what we now know about the processes of generation and recombination. It is the so-called *method of band diagrams*. That method or "language" of band diagrams is widely used when analyzing semiconductor devices. It enables describing most complicated situations in a distinct and concise form and is sometimes quite indispensable. In this section, we use the band diagram language to render the most simple notions.

In order to have the electron and hole appear in a semiconductor, it is necessary to expend the energy E_g. We will depict the process of the birth of this pair schematically (Fig. 4). Let us assume that while the electron is in the orbit connecting the silicon atoms, it possesses the energy E_v. (The index "v" denotes valence, for we deal here with valence electrons.)

In order to get a conduction electron and hole (to form an electron-hole pair), it is necessary to expend energy, or, in other words, to overcome an energy barrier whose height is E_g. The energy E_g can be obtained from thermal fluctuations of the lattice or else from a rather energetic quantum of light-photon. In case the energy barrier has been overcome, the electron gets

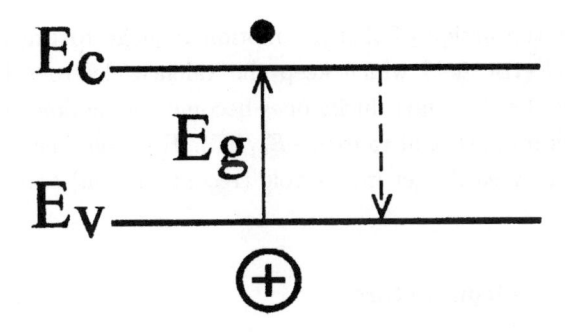

Fig. 4. Band diagram illustrating generation and recombination processes in an intrinsic semiconductor.

to the level E_c. (The index "c" denotes conductivity). So there appears a conduction electron (black dot) and a hole (white circle with the mark "+"). The band diagram makes the notion of the energy barrier distinct and obvious: it is a portion of energy, necessary for some process to take place.

The process of the recombination of electron and hole in the intrinsic semiconductor can also be easily imagined by means of a band diagram. In Fig. 4, this process is indicated by a dotted arrow. The same amount of energy E_g that was spent to form an electron-hole pair is given off during the process of recombination. Sometimes recombination is followed by the birth of a quantum of light — the photon with the energy E_g. But quite often the electron and hole recombine without giving birth to any photon. Then the energy E_g is transferred to the lattice, heating it up. On our request, the artist represented the band diagram in Fig. 5 even more vividly. Only very few

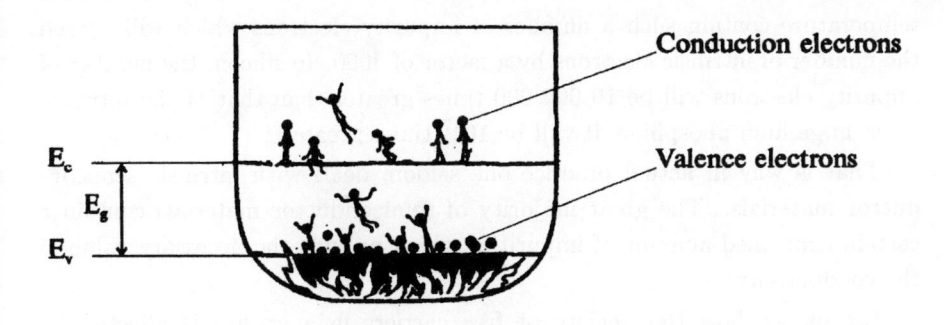

Fig. 5. Vivid illustration of the band diagram in Fig. 4.

electrons under the action of thermal motion manage to break through the fetters of the electric field which keep the valence electrons in interatomic orbits (at the E_v level). Those lucky ones become conduction electrons. Their concentration is proportional to $\exp(-E_g/kT)$. However, they should also be vigilant. They may easily get into a hole (recombination) and end up on the E_v level.

1.2 Impurity Semiconductors

There are no ideal things in nature. And there are no substances containing atoms only of one definite kind. Any real crystal, including the semiconductor crystal, always contains certain foreign impurities. Impurities are found in natural raw materials from which semiconductors are synthesized. Besides, there are impurities in the walls of the furnaces and installations where the synthesis and purification of semiconductors take place and in the atmosphere as well.

One cubic centimetre of various crystals contains $\sim 10^{22}$ atoms. A substance is usually called pure if it contains one foreign atom per 1000 intrinsic atoms (i.e. the impurity concentration is $\sim 0.1\%$). From the chemical point of view, the substance will be absolutely pure if it contains one foreign atom per 100 000 intrinsic atoms. Nevertheless it means that every cubic centimetre of the substance will contain $\sim 10^{17}$ foreign atoms. And now let us imagine that there is an impurity in the semiconductor which is able to release quite easily a free electron or form a hole. (Later on, we will see that very many impurities possess this property). Then germanium, absolutely pure from the chemical point of view, (the concentration of impurity will be only 0.001%) will at room temperature contain such a number of impurity electrons which will exceed the number of intrinsic electrons by a factor of 4000. In silicon, the number of impurity electrons will be 10 000 000 times greater than that of the intrinsic ones: in gallium phosphide, it will be 10^{17} times greater!

That is why in actual practice one seldom deals with intrinsic semiconductor materials. The great majority of semiconductor materials contain a certain controlled amount of impurities which provide the necessary value of the conductivity.

Let us see how the density of free carriers in a crystal is affected by impurities. We will begin with the simplest example.

1.2.1 Donor impurity

Let us assume that an intrinsic atom has somehow got into a silicon crystal and occupied a place in one of the sites of the crystal lattice, substituting its lawful host — the silicon atom. Now let us look at Fig. 6. The atom of As takes the place of the silicon atom in site 15. The silicon atom has four valence electrons; the arsenic atoms have five valence electrons. The four valence electrons of arsenic atoms make bonds with the neighbouring silicon atoms (atoms 9, 14, 16 and 21). Now what about the fifth electron?... The fifth electron will be held by the As atom, but much more weakly than the other four electrons, which are very tightly bound in their electron orbit, determined by the structure of the silicon crystal. The energy ΔE, necessary to break the bond of the fifth electron with the arsenic atom and to transform it into a free electron, is much smaller than the energy E_g necessary to break the bond between the silicon atoms and form an electron-hole pair.

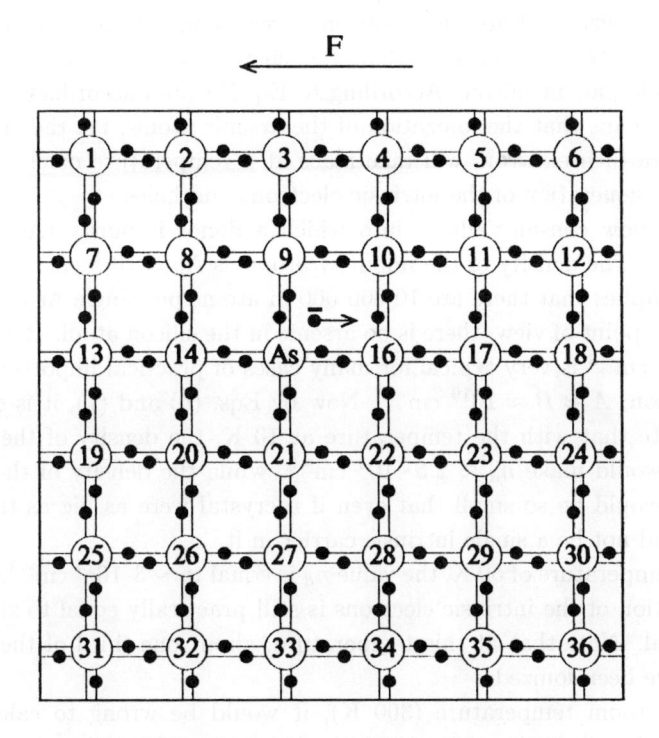

Fig. 6. The donor atom (As) in the silicon lattice.

The impurity whose atoms give away their electrons easily is called a *donor impurity*. The Latin word "*donare*" from which the word "donor" comes, means "to endow," "to give."

Let N_d atoms of a donor, say of arsenic, be introduced into every cubic centimetre of a crystal. Now let us first consider the simplest situation — the temperature of the crystal is $T = 0$ K. It is clear that in this case the crystal remains an ideal dielectric: though it takes a very small amount of energy to break away the fifth electron from the arsenic atom — the temperature being equal to absolute zero — there is no energy of thermal oscillations whatsoever.

If the temperature of the crystal is $T > 0$, then the equilibrium density of the impurity electrons n_d is determined by the expression, analogical to Eq. (6)

$$n_d = (A \cdot N_d)^{1/2} e^{-\frac{\Delta E}{kT}} \tag{7}$$

Instead of a large value E_g, we have in the exponent a much smaller value ΔE. The ionization energy ΔE (which is sometimes called the activation energy of impurity) of arsenic in silicon is very small. It is equal to 0.05 eV, which is twenty times smaller than the energy E_g necessary to create an electron-hole pair in silicon. According to Eq. (7) (and according to common sense), it means that the ionization of the arsenic atoms, i.e. the detachment of the extra fifth electron, will take place at a temperature much lower than that of the generation of the intrinsic electrons and holes.

Let us now consider silicon into which a donor impurity (As) has been introduced. The density of the impurity atoms is $N_d = 10^{15}$ cm^{-3}.

This implies that there are 10 000 000 Si atoms per single As atom. From a chemist's point of view, there is no arsenic in the silicon at all. But the value $N_d \approx 10^{15}$ cm^{-3} is very typical for many cases of practical importance.

In silicon, $A \approx B \approx 10^{19}$ cm^{-3}. Now by Eqs. (6) and (7), it is quite easy to calculate that with the temperature at 10 K, the density of the impurity electrons would make $n_d \approx 2.5 \cdot 10^4$ cm^{-3}, while the density of the intrinsic electrons would be so small that even if a crystal were as big as the galaxy, there would not be a single intrinsic carrier in it.

At a temperature of 50 K, the value n_d is equal to $\sim 3 \cdot 10^{14}$ cm^{-3}, while the concentration of the intrinsic electrons is still practically equal to zero in any real crystal. Note that at this temperature, about one-third of the impurity atoms have been ionized.

As for room temperature (300 K), it would be wrong to calculate the concentration of the impurity electrons by means of Eq. (7) since it is not exact but just an approximation. It can be used only in those cases where the calculated value n_d is much smaller than the concentration of the impurity atoms N_d, i.e. when $kT \ll \Delta E$.

But if the temperature is so high that $kT \sim \Delta E$ (or moreover if $kT > \Delta E$), then all of the impurity atoms prove to be ionized and the concentration of the impurity electrons is just equal to N_d. This phenomenon is called the "impurity saturation (exhaustion)" — the term seems quite reasonable. When the temperature is high enough, all the impurity atoms give away "extra" electrons and the source of the electrons is "exhausted." No further rise in the temperature of the crystal will increase the concentration of the impurity electrons.

In Si, room temperature (300 K) corresponds to the impurity saturation region. So in our example when $T = 300$ K, $n_d = N_d = 10^{15}$ cm^{-3}. The concentration of the intrinsic electrons n_i at 300 K is equal to only $\sim 10^{10}$ cm^{-3} (see Table 1), which is 100 000 times smaller. And only at the temperature of ~ 600 K (300°C) will the intrinsic concentration in silicon be equal to $\sim 10^{15}$ cm^{-3} and become equal to the concentration of the impurity.

When the concentration of the impurity $N_d \cong 10^{18}$ cm^{-3}, the silicon crystal should be heated up to the temperature somewhat more than 1000 K (700°C) so that the intrinsic concentration might be equal to the impurity concentration.

1.2.2 Acceptor impurity

Figure 7 shows a crystal lattice of silicon in which one of the sites is occupied by the boron impurity atom. Boron (B) is trivalent — there are three electrons on its outer electron shell. It is one electron short of making a complete bond with the neighboring silicon atoms.

Compare the picture of the electron bonds of the B atom in Fig. 7 with the configuration of the electron bonds around atom 19 in Fig. 3. The situations are quite similar, aren't they? Both the silicon atom 19 in Fig. 3 and the B atom in Fig. 7 are short of one electron. But there is also a great difference between them. All the silicon atoms are quite identical, and the empty link, the hole, which now belongs to atom 19 may at any moment approach atom 14, then atom 9 and so on.

It does not take any energy for the hole to travel in the crystal. But a boron atom is a stranger in the silicon lattice. To make the electron of the neighboring silicon atom to go over to boron, it is necessary to expend energy

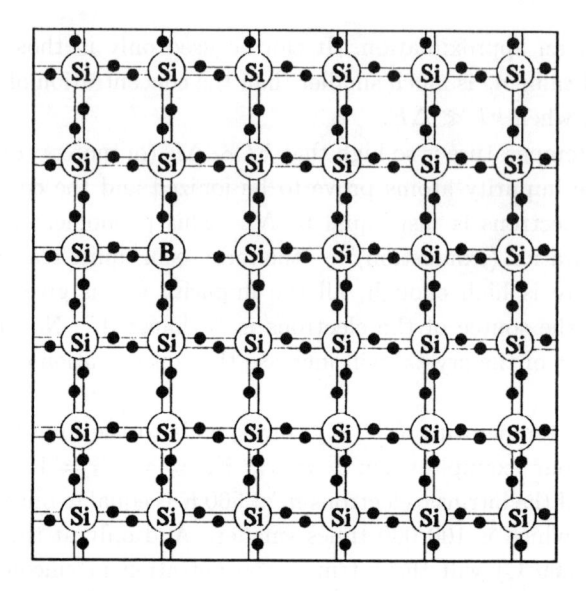

Fig. 7. The acceptor atom (B) in the silicon lattice.

ΔE. This energy, the activation energy, is not large (for B atoms in Si, it is only 0.045 eV), and yet it is not zero, as in the case of a hole in Fig. 3. No hole will be formed in the crystal until the "energy barrier" ΔE is overcome, no matter how small it is.

Let us assume that either the lattice vibrations or a quantum of light have supplied the necessary energy, and that the electron from the neighboring Si atom has come over to boron. Now the situation will be identical to that shown in Fig. 3. There is an empty link in the lattice, i.e. a hole — a free carrier of positive charge. So, acceptor impurities create holes — free carriers in a semiconductor crystal. If the temperature of the crystals is $T > 0$, then the equilibrium density of the impurity holes p_a can be found from the expression analogous to Eq. (7).

$$p_a = (B \cdot N_a)^{1/2} e^{-\frac{\Delta E}{kT}} \qquad (8)$$

The value B here is the same as in Eq. (6). N_a is the density of the acceptors.

The appearance of a hole in an impurity semiconductor is not accompanied by the appearance of a conduction electron, which is clearly seen from Fig. 7.

Equation (8), like Eq. (7), is rather approximate. It can be used only if the value p_a, calculated by it, is much smaller than the concentration of the impurity atoms N_a, i.e. when $kT \ll \Delta E$. If $kT \sim \Delta E$, or moreover, if $kT > \Delta E$, then all the boron atoms will take away electrons from the neighbouring silicon atoms, and the concentration of holes p_a will be equal to the concentration of the introduced impurity N_a.

1.2.3 *The temperature dependence of the carrier concentration*

A semiconductor into which some donor impurity has been introduced is called an electronic semiconductor, or a semiconductor of n-type. The letter "n" comes from the word "negative", showing that the semiconductor contains many negatively charged particles — electrons.

A semiconductor into which some acceptor impurity has been introduced is called a hole semiconductor, or a semiconductor of p-type. The letter "p" comes from the word "positive", showing that the semiconductor contains positively charged particles — holes.

Look at Fig. 8. The curve, shown there, gives a short summary of what we have learnt about the properties of both intrinsic and impurity semiconductors. The curve represents a typical temperature dependence of the equilibrium

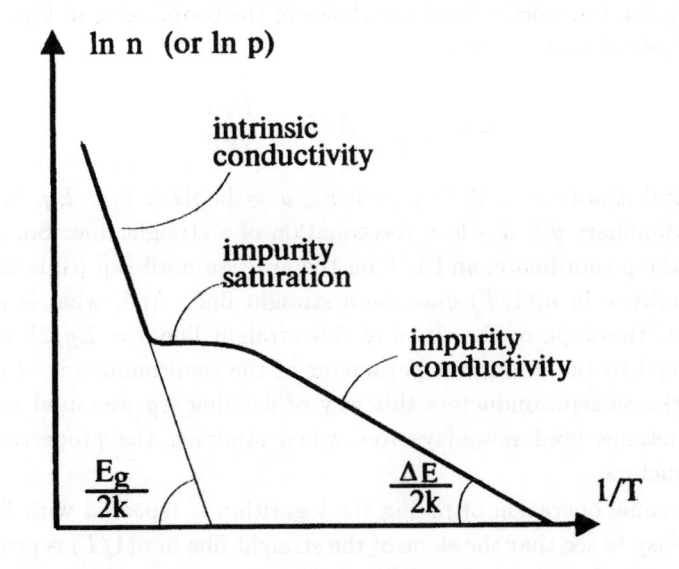

Fig. 8. The typical temperature dependence of carrier concentration.

density of free carriers in a semiconductor. The value of the natural logarithm of the concentration of electrons $\ln n$ (if the n-type semiconductor is meant) or $\ln p$ (if the p-type semiconductor is meant) is plotted on the y-coordinate. Note that it is not the value of T, but its inverse value $1/T$ that is plotted on the abscissa.

And now we will discuss the dependence shown in Fig. 8. We begin from the right, from the low temperature region, where the value $1/T$ will be great, and we will advance on the abscissa from the right to the left. As we have seen, at low temperatures the carrier (electron or hole) density in a semiconductor is determined by the density of impurity centres. The carrier density increases with the rise of temperature, and this section of the curve is determined by Eq. (7) for the electron or by Eq. (8) for the hole semiconductor. At a certain temperature this dependence is saturated. It is the impurity saturation region. All the impurity atoms have already been ionized and the carrier concentration is equal to the donor concentration (for the n-type semiconductor) or to the acceptor concentration (for the p-type semiconductor), but the intrinsic carrier density is still much smaller than that of the impurity. Finally, in the region of a still higher temperature, there is an abrupt increase of the density with a further rise of temperature. It is the region of intrinsic conductivity where the $n(T)$ dependence is expressed by Eq. (6).

And now a few words about the choice of the coordinates in Fig. 8. Let us take the logarithm of Eq. (6)

$$\ln n_i = \frac{1}{2} \ln(A \cdot B) - \frac{E_s}{2k} \frac{1}{T} \qquad (9)$$

and we will denote $x = 1/T$; $y = \ln n_i$; $a = \ln(AB)$, $b = Eg/2k$, Eq. (9) will look familiar: $y = a - bx$ - the equation of a straight line. So, if we plot $\ln n_i$ on the y-coordinate, and $1/T$ on the abscissa until Eq. (6) is valid, then the dependence $\ln n_i(1/T)$ must be a straight line. And, what is especially important, the slope of the curve of this straight line $b = Eg/2k$ is directly proportional to the important parameter of the semiconductor - Eg. In the early works on semiconductors this way of defining Eg was used very often. It is sometimes used nowadays too, when studying the properties of new semiconductors.

If the same operation of taking the logarithm is repeated with Eq. (7) or (8), it is easy to see that the slope of the straight line $\ln n(1/T)$ is proportional to the value ΔE.

1.2.4 *Minority carriers*

And now we will discuss a very important question regarding the concentration of holes in an electronic semiconductor and the concentration of electrons in a hole semiconductor.

First of all it should be mentioned that when the donor atom gets ionized and yields an electron, no hole is formed. The arsenic atom in Si lattice, for instance, having yielded the fifth valence electron remains bound with the adjacent atoms of silicon by four "valid" bonds. There is no empty place where electrons from the adjacent orbit could be displaced. Therefore no hole is formed.

On the face of it, it might seem that the electronic semiconductor must then have as many holes as the absolutely pure intrinsic semiconductor. But that view would be wrong. It is true that at a given temperature T the number of broken electron bonds and electron-hole pairs, appearing every second in an electronic semiconductor, is exactly equal to those in an intrinsic semiconductor. But the holes perish much more often, because there are more free electrons in the electronic semiconductor than in the intrinsic one. That means that a hole will meet with an electron much more often and each time it will result in a recombination and the disappearance of the hole. A a result of it, there are fewer holes in the n-type semiconductor than in an intrinsic semiconductor. There is a very simple equation connecting the equilibrium density of holes p and the electrons n in a semiconductor

$$p \cdot n = n_i^2(T). \tag{10}$$

The greater the number of electrons, the smaller the number of holes and vice versa.

The quantity n_i in Eq. (10) represents the value of the intrinsic carrier density at a given temperature T and is defined by Eq. (6).

By using Eqs. (10) and (6) it is easy to calculate that in n-type silicon, say, at room temperature, and with the electronic density of 10^{15} cm^{-3}, the hole density will be $\sim 10^5$ cm^{-3}, i.e. ten billion times smaller than that of the electrons!

So it seems quite natural that the electrons in the n-type semiconductor are called the majority carriers while the holes are called the minority carriers. In the semiconductor doped with acceptors; in the hole semiconductor, it is just the opposite — the holes will be the majority carriers, while the electrons will be the minority carriers.

The appearance of a hole in a p-type semiconductor is not accompanied by the appearance of a conduction electron, which is clearly seen in Fig. 7. The electrons in a hole semiconductor appear only on account of the generation of electron-hole pairs in exactly the same way as in the intrinsic semiconductor. But they perish much sooner than in the intrinsic semiconductor because the possibility of a collision with a hole and recombination here is much higher.

So, what happens to the electrons in a hole semiconductor is absolutely the same as to holes in the n-type semiconductor. No wonder that Eq. (10) is valid for the p-type semiconductors, as well. But now there will be more holes than electrons. Say, if in the p-type silicon the concentration of major carriers, i.e. holes, makes 10^{15} cm^{-3}, the concentration of minor carriers, i.e. electrons, will be $\sim 10^5$ cm^{-3}.

1.2.5 Band diagrams

In the previous section we discussed the main characteristics of impurities whose activation energy ΔE is not large, and makes up but a small portion of the energy E_g of generating an electron-hole pair. Such impurities are called shallow. Figure 9 indicates band diagrams, illustrating the behaviour of shallow impurities. Figure 9(a) shows the processes in a crystal which contains shallow acceptor centres. The E_a energy level differs from the E_v energy level by the value of the acceptor level activation energy ΔE_a.

With $T = 0$, the E_a level is empty, with no electrons there. Recall, "this energy, the activation energy is not large (for the boron atoms in Si, it

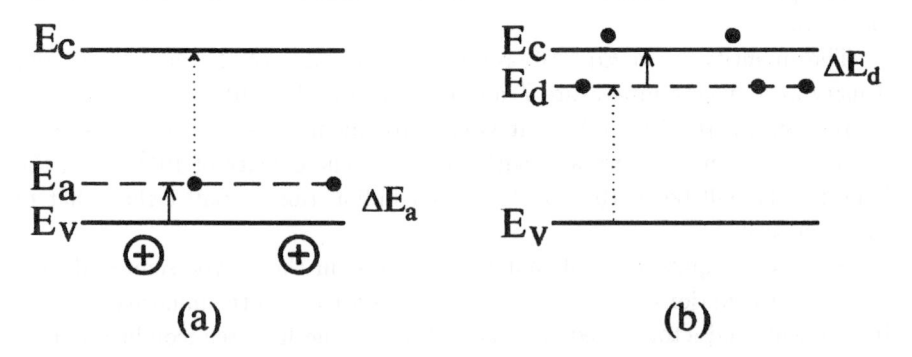

Fig. 9. Appearance of holes in the crystal with shallow acceptor levels (a); Appearance of electrons in the crystal with shallow donor levels (b).

is only 0.045 eV), and yet it is not zero. No hole will be formed in the crystal until the "energy barrier" ΔE has been overcome, no matter how small it is ..." With the temperature $T > 0$, though low enough such that $kT \ll \Delta E_a$, holes will appear in the crystal. Their concentration is defined by Eq. (8). But Eq. (8) also defines the concentration of electrons at the E_a level at the temperature T. Indeed, holes in the p-type crystal are formed just because electrons from the E_v level are captured by the acceptor level E_a. The number of electrons captured by the acceptor level is equal to the number of mobile holes which appear in the crystal.

The process of the formation of holes and of capturing the electrons by the acceptor centres is shown in Fig. 9(a) by a solid arrow. The acceptor atom which has captured the electron is charged negatively. This bound electron, unable to transfer the current, is held by the acceptor atom very tightly. It is seen from Fig. 9(a) that to release the electron, captured at the acceptor level, it is necessary to exert energy $E = E_c - E_a = /E_g - \Delta E_a/$. (The process of releasing the electron — its transfer to the E_c level, is shown at Fig. 9(a) by a dotted arrow (cf. Fig. 4)).

If $\Delta E_a \ll E_g$ (shallow level), then the energy necessary to release the electron, captured by the acceptor, is almost as large as the energy E_g.

When the value kT is of the same order as the value ΔE_a, or even greater, all the acceptor centres will be taken by electrons, and the hole concentration will be equal to N_a — the concentration of the acceptor atoms introduced into the crystal (the impurity exhaustion).

Figure 9(b) shows a band diagram of an electronic semiconductor containing the donor shallow impurity. With $T = 0$, the concentration of conduction electrons at the E_c level is equal to zero. The energy level E_d, corresponding to the donor centre, is filled up with electrons. Then the donor atoms which keep the electrons are neutral. When T is not too high, the concentration of electrons at the E_c level is defined by Eq. (7). The donor atoms which have given their electrons to the E_c level are charged positively. The concentration of those positively charged donor centres is, naturally, also defined by Eq. 7.

With $kT \geq \Delta E_d$, the concentration of free electrons at the E_c level and the concentration of the positively charged donor centres are equal to N_d (the donor exhaustion).

The positively charged donor centre can capture an electron from the E_v level. Then there appears a hole in the crystal. But in order to transfer an electron from the E_v level to the donor level E_d and thus create a hole in the crystal (this process is shown in Fig. 9(b) by a dotted arrow), it is necessary to overcome an energy barrier whose height is $E_g - \Delta E_d \approx E_g$.

Pay attention to a very important feature, which is clearly seen in the diagrams: if the electrons are not affected by any exterior disturbances, they are at the lowest of the energy levels.

Indeed, in the intrinsic semiconductor when $T = 0$ and there is no illumination, as we know there are neither electrons nor holes there. In the language of band diagrams (Fig. 4) that means that all electrons are at the lowest of all the possible energy levels, at the level E_v.

In a semiconductor containing the acceptor centres with $T = 0$ and without any illumination, all the electrons are also on the level E_v. The levels E_a and E_c are empty [Fig. 9(a)].

In the semiconductor containing donors [Fig. 9(b)], with $T = 0$, electrons are on the level E_d. But why not on the level E_v?

The thing is that the electrons tend to occupy the lowest possible energy level. In the semiconductor crystal containing donors, with $T = 0$, the level E_v is wholly filled with electrons. There are no vacant places there. And the electrons occupy the lowest possible level — the E_d level. Mind what that means in the physical sense. The level E_v corresponds to the energy of electrons binding the silicon atoms to each other. With $T = 0$, all the electrons are in their orbits (Fig. 2) and there are no vacant places (holes) which the electrons from the donor level E_d might have occupied.

Compensation. Having learnt a few phrases in the language of the band diagram, we can solve a problem in a matter of seconds, in a distinct and memorable form. To get that answer in a traditional way, one should use a rather subtle physical reasoning.

The problem is formulated as follows: let us assume that the same amount of the donor and acceptor shallow impurities are simultaneously introduced into a semiconductor. For instance, the same quantity of arsenic and boron are introduced into silicon. What will happen? The natural answer is that there will be a great concentration of free carriers of both signs — electrons and holes.

But that answer is absolutely wrong. The right answer is that there will neither be impurity electrons nor impurity holes in the semiconductor, the latter behaving as an intrinsic semiconductor. Look at Fig. 10(a). All the electrons from the donor level E_d moved over to the lowest initially vacant acceptor level E_a. Now, in order to create a free electron in the crystal, it is necessary to spend the energy $E_g - \Delta E_a$, which is almost as large as the energy E_g, necessary to create an electron-hole pair in an intrinsic semiconductor. It is hardly possible for the hole to appear in the crystal: all the vacant places on the acceptor level have been filled up by the electrons from the donor level E_d. There are no places on the acceptor level for the electrons from the E_v level. Since there might appear a hole in the crystal, it is necessary either to transfer an electron from the E_v level directly to the E_c level (which will require the energy E_g), or to vacate a place on the E_a level. To do it, it is necessary to transfer the electron from the E_a level to the E_c level (the energy needed for $E_g - \Delta E_a$ is almost as large as E_g). It is enough to cast a look at Fig. 10(a) in order to see that the semiconductor, into which the donor and acceptor shallow impurities were introduced in the same concentration will behave almost in the same way as the intrinsic semiconductor. This phenomenon is called *compensation*. Introduction of the acceptor compensates for the action of the donor impurity and vice versa.

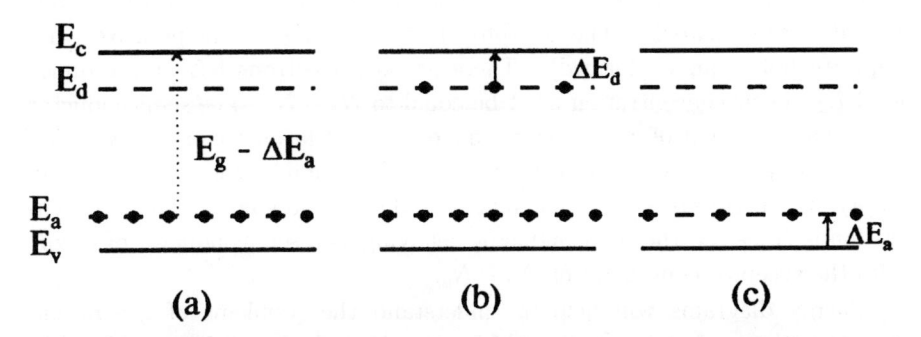

Fig. 10. Compensation. (a) Full compensation. The donor concentration N_d is equal to the acceptor concentration N_a. (b), (c) Partial compensation: $N_d > N_a$ (b); $N_d < N_a$ (c).

To what physical picture does the energy diagram in Fig. 10(a) correspond? Let us reason it out. Atoms of the donor impurity introduced into the crystal will easily give away their redundant electrons which will become free. Atoms of the acceptor impurity will capture electrons from their nearest lattice neighbours and will form an adequate number of holes. The impurity electrons and holes will begin wandering chaotically about the crystal and sooner or later will meet and recombine. In accordance with the conditions of the problem under consideration, the number of electrons which can easily be released is exactly the same as the number of holes (since the concentration of shallow donors is equal to that of the acceptors). After the impurity electrons and holes have met, recombined and disappeared, there will be no more atoms in the crystal which might easily generate an electron or hole.

Indeed, if we consider the example with the silicon into which arsenic and boron were introduced, we will see that for each arsenic atom there will be only four valence electrons left, held almost as securely as the four valence electrons held by the silicon atom. The same can be said of the boron atom. On gaining the forth electron, the boron atom formed bonds with the neighboring Si atoms as solid as those between the silicon atoms. As a result of that, in order to generate a conduction electron or hole, it is now necessary to spend an amount of energy almost as large as that spent on generating an electron-hole pair in the intrinsic silicon.

And what will happen if the concentrations of the donor and acceptor impurities are not equal? Figure 10(b) and (c) allows us to answer that question.

Figure 10(b) refers to the case when the amount of the donor impurity is greater than that of the acceptor impurity, and the donor impurity is but *partially compensated.* The acceptor level is "rendered inoperative" (no impurity holes can be formed). There are still electrons left at the donor level E_d. Their concentration must be equal to $N_d - N_a$. The semiconductor will behave like that of n-type, whose donor concentration is equal to $N_d - N_a$. Figure 10(c) illustrates the case when $N_a > N_d$. The donor level is absolutely empty, and there are still vacant places at the acceptor level. Their concentration is $N_a - N_d$, the semiconductor behaving as a semiconductor of p-type with the acceptor concentration $N_a - N_d$.

Energy diagrams will help us understand the problems of generation, recombination and the movement of free carriers which will be considered in the next chapter. But the full meaning of the diagrams will become clear only in the second part of the book, where we will study the energy barriers, which are formed on the boundaries between the semiconductor and the surrounding medium or on the border between the semiconductors of n- and p-type.

1.3 Deep Levels

There is a long list of impurities for any semiconductor which is commonly used, such as silicon, germanium, gallium arsenide, indium phosphide, etc. For silicon, for instance, this list contains several dozen names. It includes such familiar substances as arsenic, boron, phosphorus, aluminum, silver, copper, cadmium, cobalt, gold, iron, oxygen, mercury, platinum, molibden, nickel, palladium, sulphur, selenium, tungsten, zinc and many other elements. Each of those elements is characterised, as an impurity in silicon, by its activation energy ΔE, the values of those energies lying in the range from 0.05 eV (shallow donors and acceptors) up to 0.8-0.9 eV. The latter values, characteristic of the "deepest" impurities, are quite comparable to the energy E_g in silicon (1.1 eV).

One might think that the "deeper" the impurity (i.e. the greater ΔE), the smaller the role it has to play. And really, it is clear from Eqs. (7) and (8) that with the increase of ΔE the density of free electrons (or holes) falls very fast exponentially. Equations (7) and (8) are valid for any value of ΔE. But the conclusion that the "deep" impurities can be neglected would be quite wrong. Deep impurities in semiconductors play three very important roles: they act as "compensators", centres of generation and centres of recombination. We will consider each of these roles in turn.

1.3.1 Compensation by deep levels

Let us imagine that we want to obtain a semiconductor, say GaAs, with a very high resistivity. Specialists manufacturing the semiconductor devices are frequently facing such problems. Films or plates whose resistivity is very high are often used as a substrate to which very thin layers of the same semiconductor containing different doping impurities are applied.

The intrinsic semiconductor has the highest resistivity: its number of free carriers being the smallest. But what we have already learned about the impurities in semiconductors is quite enough to make us realize how difficult it is to obtain intrinsic semiconductors. Even the smallest shallow impurity density makes the density of electrons and holes increase by thousands, millions and even billions of times.

Every effort to get rid of shallow impurities entails long, complicated and expensive operations to purify the materials. But so far, appreciable results have been obtained only for germanium and silicon. For GaAs, for instance, no matter how perfect the modern technology is, the shallow impurity density level is still $N \geq 10^{12}$ cm^{-3}. This value is ten billion times smaller than the density of the gallium and arsenic atoms in GaAs, but it is still a hundred thousand times greater than the density of the intrinsic electrons and holes in the same material at room temperature (see Table 1). So, what is there to be done?

The answer is quite simple. We will not try to purify the semiconductor very thoroughly. Let it contain 10^{15} or even 10^{16} cm^{-3} of shallow donors. And we will also introduce into the semiconductor an additional impurity, whose concentration is still greater, say 10^{16} or 10^{17} cm^{-3}. But this second impurity is to be a deep acceptor.

What will happen here? As we know, electrons created by the shallow donors continue their chaotic wandering caused by thermal motion. Sooner or later, each of these electrons will approach an acceptor impurity atom which is always ready to trap the electron. The deep acceptors will take up all the electrons created by the shallow donors and it will not be easy to release the electrons now. The activation energy of these electrons from deep impurities ΔE is very great.

Figure 11 shows a band diagram illustrating the compensation of the donor level by a deep acceptor level. Figure 11(a), in fact, repeats Fig. 9(b). The semiconductor contains only some shallow donor impurity with ionization energy ΔE_1. The temperature T is great enough and all the electrons from the shallow donor move to the E_c level (they become free). Figure 11(b) corresponds to the situation when apart from the shallow donor, the crystal contains an element which is a deep acceptor (the E_t level). The index "t" comes from the word "trap." All the electrons from the shallow donor have been taken up to the deep acceptor level. To free them it is necessary to spend a large energy $\Delta E_2 \gg kT$. There are practically no free carriers in the crystal.

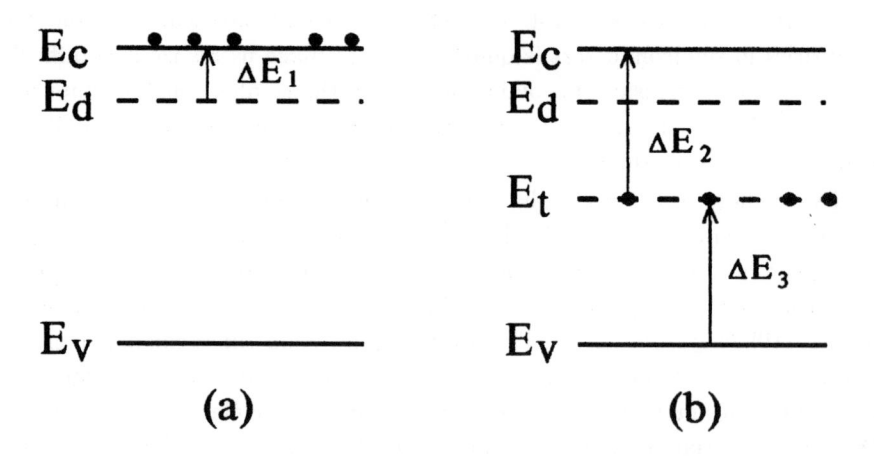

Fig. 11. Compensation with a deep impurity level.

With the acceptor impurity being introduced, the number of free electrons decreases abruptly. Thus, introducing an acceptor impurity actually compensates for the presence of the donor impurity. It would not matter at all if the amount of deep acceptors is even several times greater than that of the shallow donors. All the electrons created by shallow donors are sure to be taken up (captured). But what about the "extra" acceptors?

They are capable of taking up electrons from the semiconductor atoms and forming holes. But they are *deep*, their activation energy ΔE_3 is great and accordingly, the hole concentration will not be great at all.

In the case of GaAs, either chromium or oxygen is introduced into GaAs, which has not been very thoroughly purified. If before the incorporation of a deep impurity, GaAs was of the n-type, then chromium is introduced. Chromium is a deep acceptor in GaAs.

If the initial GaAs was of p-type, then oxygen is introduced; it is a deep donor in GaAs. Deep donors compensate for the shallow acceptor impurity just like the deep acceptors compensate for the shallow donor impurity.

This results in obtaining monocrystals of GaAs with a very high resistivity of up to 10^7 Ohm cm. The substrates of the "semi-insulating" gallium arsenide are widely used when manufacturing GaAs devices.

1.3.2 *Generation through deep levels*

We know that it takes the energy E_g for the appearance of an electron and hole in an intrinsic semiconductor. Figure 12(a) gives again (compare with Fig. 4) a schematic diagram illustrating the process of electron-hole pair generation.

If the energy E_g is to be obtained from the thermal fluctuation of the lattice, and the value E_g is much greater than the mean thermal energy kT, then this process is hardly probable. As we know, it is proportional to $\exp(-E_g/kT)$ [see Eq. (2)]. But is it possible to expend the energy necessary to release the electron not all at once but in two steps, as shown in Fig. 12(b)? First, impart the first portion of the energy, i.e. $E_g/2$, and then the second portion, also equal to $E_g/2$. It's easy to see that if we managed to do this, the process of generating electrons and holes would be much more intensive.

Perhaps the following comparison might illustrate this idea. Let us imagine that it is we and not the electron who must overcome a high barrier at one go without any stairs. If the barrier is two metres high, then among the several billion people inhabiting our planet, only a few dozen will be able to overcome it. But the same barrier was divided into two parts, each one metre high, and

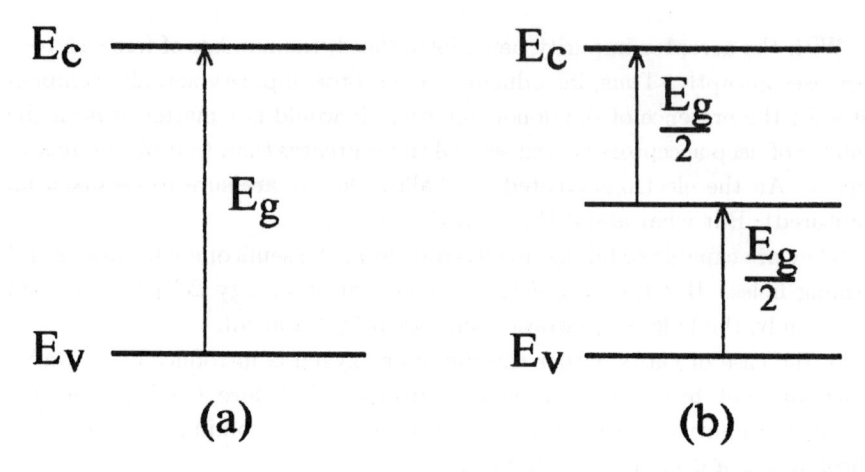

Fig. 12. Schematic diagram of electron-hole pair generation. (a) the barrier is being overcome at one go; (b) the same barrier is being overcome in two steps.

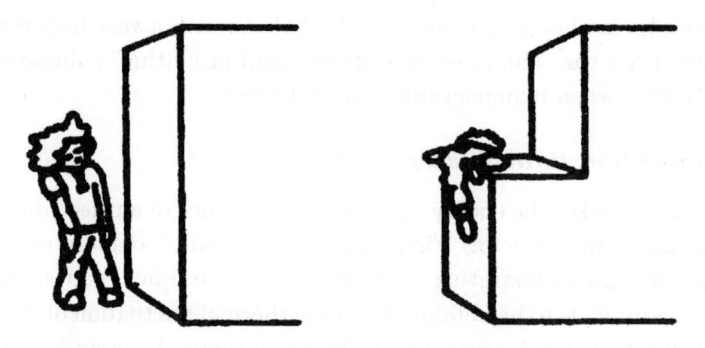

Fig. 13. It is much easier to ascend two stairs, one metre high each, than to overcome a barrier two metres high at one go.

could be easily overcome by many more people. Perhaps every tenth man would be able to do it (Fig. 13).

We will now see that the deep levels whose activation energy is $E_g/2$ are in fact those "stairs" — "levels" — shown in Fig. 12(b).

Let us assume that the acceptor impurity centres whose activation energy is $\Delta E = E_g/2$ are introduced into the semiconductor. The acceptors will take away the electrons from the neighbouring atoms of the semiconductor. Thus the holes formed will begin moving chaotically in the crystal. Earlier, when we examined the behaviour of the shallow acceptor, we did not think of the electron taken by the acceptor atom. And that was right. The small activation energy ΔE corresponds to a very small step up the energy stairs [Fig. 9(a)]. Releasing the electron from the shallow acceptor and freeing it requires a lot of energy, practically equal to E_g. It is quite different now, when the electron is taken up by a deep acceptor. Thermal fluctuations of the lattice are most likely to break the electron from the acceptor and free it. As a result of such a "two-step" process, an electron-hole pair has been created in the crystal. The acceptor has again acquired an ability to take away an electron from the neighbouring atom of the semiconductor and form the next hole, etc. The generation rate of electron-hole pairs increases abruptly.

A process consisting of two stages is most effective for the impurity whose activation energy is $E_g/2$.

Let us assume a semiconductor material to have an acceptor level. If the acceptor's activation energy is less than $E_g/2$, then the electron from the level E_v will jump to it quite easily. But it will be rather hard to transfer it to level E_c (that is to set it free). Here is an exact analogy with a staircase: if you are to climb up a height of two metres and the only flight of stairs is 30 cm high, it will be quite easy to climb it. But as for the remaining 1.7 m, it would be almost as hard to climb it as the original two-metre height. The optimal version would be to have two flights of stairs of equal height.

So, introducing deep impurities greatly increases the rate of generating electron-hole pairs. It may seem that in this case, in the presence of deep levels, the concentration of electrons and holes must grow considerably. However, this is not so. In the next section, we will see that the deep levels increase not only the rate of their generation, but also the speed of the recombination of electrons and holes. As a result, Eqs. (7), (8) and (10) hold true and give correct descriptions of the equilibrium densities of electrons and holes. But in this case, does it matter if there are any deep centres in a crystal at all?

The point is that so far we have been speaking about the simplest situation, the *equilibrium* situation where the number of carriers created by thermal motion is exactly the same as that of the carriers perishing during that time due to the recombination under those conditions. The impurities cannot manifest themselves as generation centres. Due to their presence, the increase in the rate of carrier generation is absolutely the same as the increase in the rate of their own recombination.

To make the activities of the generation centres explicit, it is necessary to break the equilibrium between the processes of generation and recombination, creating *a non-equilibrium situation*. The best way to do it is by creating conditions when the process of generation will go on under the same conditions of equilibrium, while the process of recombination will not take place at all. How can we manage to do it?

It proves to be not so difficult after all. It is necessary to create a strong electric field in some part of the semiconductor. Then the free carriers — both electrons and holes — will be carried away from that region very rapidly and their density in that region will become very small. So the rate of recombination, being proportional to the product of the densities of both electrons and holes [see Eq. (3)] will diminish sharply, there being nothing to recombine. And what about the generation process? It is affected only weakly by the electric field. The electric field, able to deliver the region of a semiconductor from any free carriers, will have practically no influence on the generation rate.

The electric field, removing the free carriers, stipulates the appearance of an electric current in the circuit. The more free carriers generated in the semiconductor, the stronger the electric current. The current in the circuit is directly proportional to the generation rate of electron-hole pairs.

In many semiconductor devices, either diodes or transistors at a certain regime of work, there is a region of a strong electric field. The current flowing through this device at the given regime is determined by the densities of the deep levels introduced into the semiconductor.

1.3.3 *Recombination through deep levels*

As we have already mentioned, Eqs. (7) and (8) are valid for any levels — deep and shallow. According to these equations, the deeper the level, the smaller the equilibrium concentration of the carriers generated by that centre at the given temperature T.

On the other hand, as we have seen, the deep levels with the ionization energy $\Delta E \approx E_g/2$ increase abruptly the rate of generating the carriers. How can both these conditions be realised simultaneously?

That is possible in but one case: if the deep centres, which abruptly accelerate the generation of electron-hole pairs, does not abruptly accelerate the annihilation (recombination) of these pairs less.

As we saw in the previous section, for the levels to act as generation centres, there must be a non-equilibrium situation in the crystal. The carrier concentration in the crystal ought to be smaller than the equilibrium concentration. For the levels to act as recombination centres, a non-equilibrium situation is also required. But in this case the carrier concentration in the crystal must be greater than the equilibrium concentration.

Let us illuminate a specimen of the semiconductor material we are investigating with the light whose quantum energy $E_{ph} = h\nu$ is greater than the energy of creating an electron-hole pair in the material E_g. There will be excess electrons and holes in the specimen (compared to those in equilibrium) and the specimen conductance will increase.

We switch off the light. It is clear that soon the density of electrons and holes will reach the equilibrium value. The excess electrons and holes recombine, "die out". Why?

The answer is clear: after the light has been switched off, the balance between the generation and recombination is broken. The generation process will create a certain number of carriers per unit of time. In the state of equilibrium, the number of new born carriers is the same as that of the carriers perishing on account of recombination. But immediately after the light was switched off, the excess carriers were still to be found in the specimen and the number of both electrons and holes was still greater than in the state of equilibrium. The rate of recombination is proportional to the number of electrons and holes per unit volume [see Eq. (3)]. So the electrons and holes perish more quickly than they appear. The density of the carriers will get smaller. It will diminish until it reaches the state of equilibrium and then the rate of generation will again become equal to the rate of recombination.

The time it takes the density to return to the state of equilibrium is determined by the *lifetime* τ of the excess carriers. The sooner the density returns to its equilibrium value, the shorter the lifetime of carriers, and the faster the recombination process.

Measuring the lifetime τ in specimens with different impurities, we can study the influence of various impurity centres on the recombination rate.

Introducing shallow impurity centres will increase the carriers' equilibrium density in the specimen but will not influence their lifetime.

But introducing deep impurities whose activation energy ΔE approaches $E_g/2$ will immediately expose them as "killers." Even though their density is very small, these impurities can shorten the lifetime hundreds of times. Thus introducing gold or platinum in a concentration of $\sim 10^{13}$ cm^{-3} (0.000 000 1%) into silicon accelerates the death of the non-equilibrium carriers by one thousand times: their lifetime in silicon shortens from $\sim 10^{-3}$ up to 10^{-6} s. (For gold and platinum, the values ΔE in Si are equal to $E_g/2$).

Then why do the deep levels act as effective centres of recombination?

Let us imagine an electron and hole wandering in the crystal. In order to meet, recombine and disappear, it is necessary that they should be close to each other in the vicinity of one and the same atom of the crystal lattice. Such a situation is, in general, possible but seldom occurs.

Let us now assume that there is an impurity centre in the crystal whose activation energy ΔE is great. Should an electron appear in the vicinity of this centre, it is sure to be trapped by the impurity centre.

The centre will keep the electron trapped until a hole appears in the vicinity. As soon as that happens the electron and hole recombine. The "killer" has committed this task of annihilating the electron-hole pair, and is ready to start all over again.

Sometimes it is important for electrons and holes to perish in the device as soon as possible. It is often quite essential for the fast switching of semiconductor devices. Then impurities creating effective recombination centres should be incorporated into the material. Sometimes, on the contrary, electrons and holes must live long. In this case, the semiconductor is to be thoroughly purified.

1.4 Summary

Semiconductors are such materials whose conductivity at the temperature of absolute zero is equal to zero, all valence electrons being bound in the interatomic orbits and thus unable to conduct an electric current.

The energy of the electron bonds is not very great, and even when the temperature is not high, the electron bonds break down on account of thermal motion. There appear free electrons (conduction electrons) and holes. Their concentration increases exponentially with the increase of temperature. And accordingly with the rise of temperature, the conductivity of semiconductors increases exponentially too.

Some impurities (donors) are able to give away their electrons quite easily, while others (acceptors) can take electrons from the semiconductor atoms. Introducing such impurities, shallow impurities, even in very small quantities, may increase the number of free carriers to a very great extent and thus greatly increase the semiconductor conductivity.

Introducing into a semiconductor deep levels with a great ionization energy ΔE increases both the generation rate and the recombination rate of electrons and holes. With $\Delta E \approx E_g/2$ the increase of the rates of generation and recombination is especially great. The influence of the deep levels is especially strong in non-equilibrium situations. When the concentration of electrons and holes is greater than under the equilibrium conditions, the presence of deep centres is displayed by an abrupt decrease of the lifetime of non-equilibrium carriers τ. When the concentration of electrons and holes is smaller than the equilibrium concentration (say, under the action of the electric field), the presence of deep centres is displayed by an abrupt increase of the current.

Chapter 2

Motion of Electrons and Holes inside the Crystal

> Das Wandern ist des Mullers Lust,
> das Wandern...*
>
> Franz Shubert,
> The words by V. Muller
> from a series of songs
> *"The Miller's Fair Wife."*

Atoms of any substances are in constant thermal motion. Very frequent, collisions of the free carriers with the atoms of the crystal lattice (they collide hundreds of billions, trillions of times per second) result in electrons and holes spending their lives in constant chaotic thermal motion. If an electric field is applied to a crystal, then the charge carriers without stopping their chaotic thermal motion also acquire a directed motion. The negatively charged electrons will drift to the positive electrode, while the positively charged holes will drift to the negative one. The directed motion of free carriers under the action of the electric field is an electric current.

The electric current can take place, as we will soon learn, even if no external electric field is applied to a crystal. Whenever a certain part of a crystal has more free carriers (e.g. electrons) than the adjacent region, then under the action of the chaotic thermal motion, the carriers begin their directed motion

---
*O, motion... motion is the miller's life...

towards the region of a lower density. This phenomenon is called *diffusion* and the electric current created by it is called the *diffusion current*.

In this chapter, we will consider the main types of motion of the free carriers: the thermal motion, the motion in the electric field and diffusion.

2.1 Thermal Motion

Free electrons and holes are in the state of chaotic motion under the action of the lattice thermal fluctuations.

The average energy of the thermal motion of electrons (or holes) equals $3/2\ kT$. Equating this value to the kinetic energy of the particle $mv^2/2$, we can find the average velocity v_T of the chaotic motion of the electron (or the hole):

$$v_T = \left(\frac{3kT}{m}\right)^{1/2} \qquad (11)$$

When $T = 300\ K$, assuming that the mass of the free carrier is equal to the mass of the free electron, we find that the velocity v_T is $\sim 10^5$ m/s.

2.1.1 *Energy distribution of electrons*

The average energy and average velocity of thermal motion are very important characteristics of the free carriers in a semiconductor. It is clear however that due to the random chaotic character of thermal motion at any given moment of time among the free carriers (electrons and holes), there are sure to be such whose energy and velocity are much higher than their average values.

Let us assume that the concentration of free electrons in the n-type semiconductor is equal to n_0. How many "hot" electrons whose kinetic energy is much greater that the average energy does any cubic cm of the crystal contain? Or, in other words, what is the concentration of electrons whose kinetic energy $\Delta E \gg kT$?

Read again, please, the paragraph, preceding Eq. (1)! This concentration can be determined by an approximate equation

$$n \approx n_0 e^{-\Delta E/kT} \qquad (12)$$

Or, if we are to count the free electron energy from the E_c level,

$$n \approx n_0 e^{-(E-E_c)/kT} \qquad (12a)$$

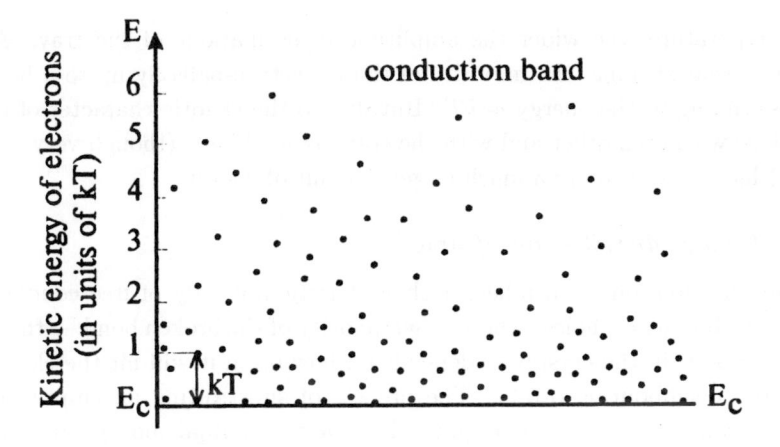

Fig. 14. Energy distribution of conduction electrons (light table tennis balls on a shaking tray). Energy is measured in kT units.

The energy distribution of electrons can be readily illustrated at the energy diagram (Fig. 14). This diagram does not look ordinary. First of all, the E_v level which is usually given at the energy diagrams together with the E_c level is not seen there.

The thing is that so far we depicted the energy diagrams of the processes whose energy by the order of magnitude was equal to $E_c - E_v = E_g$, i.e. for the typical semiconductor it attained the value of ~ 1 eV. In this case we are interested in the distribution in which the characteristic energy scale is a much smaller value of the order of kT (0.025 eV at room temperature). What happens to the electrons at the E_c level seems to be viewed by us through a microscope, the E_v level being far below, beyond the field of vision of the lens.

The kinetic energy of electrons at the E_c level is equal to zero. The increase of the kinetic energy corresponds to the upward motion of the electron in the diagram. We can see in Fig. 14 that as it follows from Eq. 12, the majority of electrons have their kinetic energy of the order of kT. There are, however, some very energetic electrons with greater energy. However, they are not numerous.

One can see in Fig. 14 that the liberated conduction electron has a vast area (zone) of energies $E > E_c$. This area of energy is called a *conduction band*.

The picture shown in Fig. 14 resembles very much the motion of plastic balls used to play ping-pong (table tennis) on a shaking tray. The higher the temperature, the wider the amplitude of oscillations of the tray. As a result of that shaking, the majority of those electron-balls fly up to a height, corresponding to the energy $\sim kT$. But due to the chaotic character of their collisions with each other and with the tray, some of balls (though very few of them) happen to acquire a much larger amount of energy.

2.1.2 Energy distribution of holes

Energy distribution is an inherent characteristic not only of free conduction electrons, but also of holes. The average velocity of the broken bond — the hole moving about in the crystal — depends on how often it will hit the electrons from the neighbouring orbits. The process of substituting an empty bond, being a random process, determined by chaotic thermal motion, it is clear that at any moment there will be holes rushing chaotically either quicker or slower than the others.

The average energy of the thermal chaotic motion of holes is, naturally, $3/2\ kT$ and the average thermal velocity is defined by Eq. (11). The energy distribution of holes is described by an approximate expression, analogous to Eq. (12):

$$p \approx p_0 \cdot e^{-\Delta E/kT} \qquad (13)$$

or, if to count the hole energy from the level E_v

$$p \approx p_0 \cdot e^{-|E-E_v|/kT} \qquad (13a)$$

An attentive reader might wonder at the sign (| |) in the exponent index. This sign shows that to make calibrations in accordance with Eq. (13a), one should use the modulus of the difference $E - E_v$. Where has this sign come from? And what does it mean?

The energy diagram, illustrating the energy distribution of holes is shown in Fig. 15. The E_c level is not shown in the diagram. The scale chosen here puts it well above E_v. The overwhelming majority of holes have a kinetic energy of the order of kT and they are within the energy level which is at the distance $\sim kT$ from the E_v level. Only a few most energetic holes manage to acquire a much larger energy. They are at the distance of kT fold from the E_v level. On the face of it, the situation is similar to that of electrons. However, as is seen from Fig. 15, there is one rather significant difference. In the energy diagram, the electron increase of energy corresponds to the motion of electrons

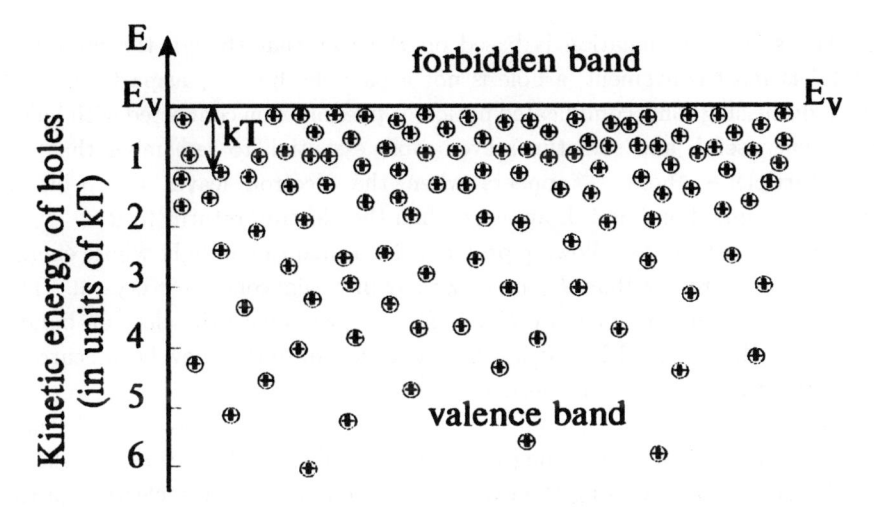

Chapter 2. Motion of Electrons and Holes inside the Crystal   43

Fig. 15. Energy distribution of holes (table tennis balls in a covered bucket filled with water).

upward along the energy scale, while the hole increase of energy corresponds to the motion of holes downward along the energy scale.

Thus, if the electrons resemble small balls on a shaking tray, then the holes are the same balls for the table tennis in a bucket of water. If the bucket is at rest, all the holes will flow up, taking their places near the E_v level just beneath the cover. If we begin to shake the bucket, some balls, as a result of that thermal motion, can sink rather deep.

Why does the hole energy increase correspond to the motion of holes downward, but not upward along the energy scale? It is not quite easy to explain this. Frankly speaking, this is one of the more difficult points (if not the most difficult one) in this book. Therefore we suggest not one but three different explanations.

The first explanation can be called "Explanation for Princes". One of the most brilliant French mathematicians of the 18th century being unable to explain the essence of some complicated phenomenon to the heir of the French throne, is said to exclaim: "Your majesty! Believe me it is so!"; to which the Royal Prince replied: "Sir! Why have we lost so much time? Word of the honest nobleman is quite enough ...". So, the first way is simply to trust and remember that the energy growth of a hole, unlike that of the electron, corresponds to the downward (and not upward) motion along the energy scale.

The second explanation is based on the fact that though the notion of "hole" is most convenient, a hole is not a particle, but a quasiparticle. And when discussing some complicated notion or phenomenon connected with holes, it is very useful, and sometimes even most essential, to remember that the quasiparticle — hole — appears when the electron leaves the orbit of the interatomic bond and disappears when the electron returns to it.

Let us consider the following problem: let a quantum of light whose energy E_{ph} is much greater than E_g be absorbed in a semiconductor crystal. The energetic photon whose energy $E_{ph} > E_g$ not just breaks the electron bonds, the rest of the energy $\Delta E = E_{ph} - E_g$ is transferred to the "new born" carriers — electron and hole — as kinetic energy.

Let us first assume that all the excess energy is transferred to the new born electron. The energy diagram, corresponding to the process, is shown in Fig. 16(a). On receiving the excess kinetic energy ΔE, the electron in the conduction band on the energy diagram flew up very high. The energy gap, i.e. the distance in the energy diagram between the electron and the hole, is equal to E_{ph}, just as it should be on account of the law of conservation of energy.

And now let us suppose just the opposite, that all the excess energy of the photon ΔE was transferred not to the electron, but to the hole. How are we to depict such a process at the energy diagram?

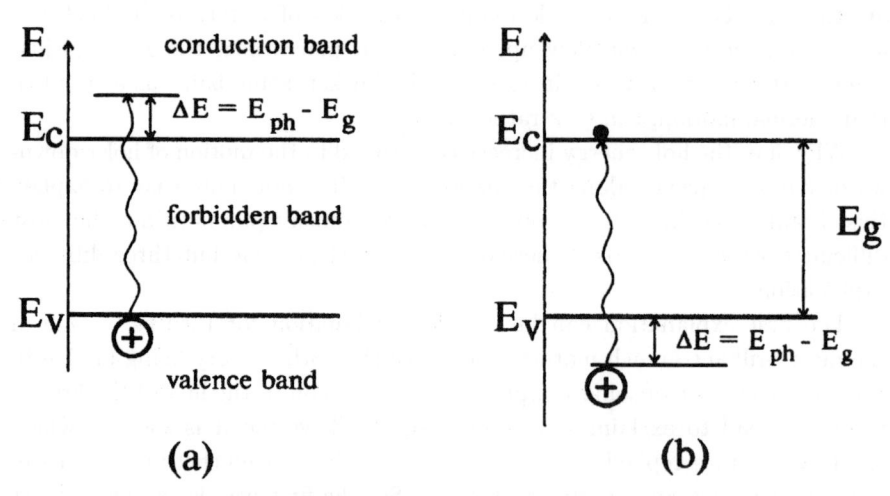

Fig. 16. The larger kinetic electron energy ΔE, the higher is the electron position at band diagram (a). The larger kinetic hole energy ΔE, the lower is the hole position at band diagram (b).

Chapter 2. Motion of Electrons and Holes inside the Crystal   45

According to the condition, the new born electron has a minimum of energy possible. So it should be placed in the diagram as close as possible to the E_c level at the lower border of the conduction band [Fig. 16(b)]. And where should we place in that diagram the energetic hole whose kinetic energy is ΔE ?

The law of conservation of energy strictly demands that the energy gap between the new born electron and hole be E_{ph}. Consequently, the hole can only be placed at $\Delta E = E_{ph} - E_g$ below the E_v level [Fig. 16(b)]. The larger the kinetic energy of the new hole ΔE, the lower it is the energy diagram.

The third explanation is meant for the most searching and inquisitive students. The law of conservation of energy is certainly not to be argued about. And still this trick of placing the hole beneath the E_v level calls for some inner protest.

First of all, this is because it is not clear what the energy band located below the E_v level corresponds to. When we introduced the notion of this level (Fig. 4 and the text referring to it), we meant the energy of an electron bond in the interatomic orbits of the silicon atoms. This is as though we took the E_v level to be the initial point of counting the energy. Indeed, if the electron at the E_v level is given a certain amount of energy, it may pass over to the acceptor level [Fig. 10(c)]. If the electron at the E_v level is imparted to the energy E_g, it may be released (Figs. 4, 5). A still larger portion of energy will make the electron energetic and "hot" with a large amount of kinetic energy [Fig. 16(a)]. With energies larger than E_v, everything is clear. But what does the energy band smaller than E_v, correspond to? To answer this rather difficult question, we must realize what "the energy of the electron, bound in the interatomic orbits of silicon atoms" is.

Imagine that there is a single isolated atom. To tear a valence electron loose from its outer electron shell it is necessary to spend a certain amount of energy, called ionization energy E_i [Fig. 17(a)]. Let us now have another atom identical to that, approach that atom. While they are far from each other, the value of the ionization energy of either of them is absolutely the same and equals E_i. However, if we bring them closer to each other so that the interaction of their electron orbits becomes appreciable, i.e. the electrons may from time to time leave one atom for another, the situation will be different. First of all, it can be seen that the energy values of the ionization of atoms will no longer be equal. In this case there are two different values of ionization energy [Fig. 17(b)]. The energy level E_i, corresponding to the valence electron, is then said to split. Secondly, neither of these new values of the energy, generally speaking, is equal to the E_i value of an isolated atom. And, thirdly, the closer the atoms are to each other and the stronger the interaction of their electron orbits, the greater, naturally, is the energy gap ΔE between the split levels.

But if there are four interacting atoms, located not far from each other, the level E_i will be split into four levels [Fig. 17(c)]. When there are N interacting atoms, the level will be split into N levels [Fig. 17(d)]. The splitting of levels when interacted is a phenomenon observed not only among atoms and electrons. It is a property of any oscillatory system in mechanics and radio engineering.

In Fig. 17(e), to the left, you can see a single circuit LC. With the capacitance C_0 and inductance L_0, the circular frequency of oscillators withinel the circuit is known to be $\omega_0 = 1/\sqrt{L_0 \cdot C_0}$. Now we can take an identical circuit with the same values of capacitance and inductance C_0 and L_0 and place it close to the first one, so that the interaction between the inductances of the circuits might be noticeable (i.e. the mutual inductance coefficient

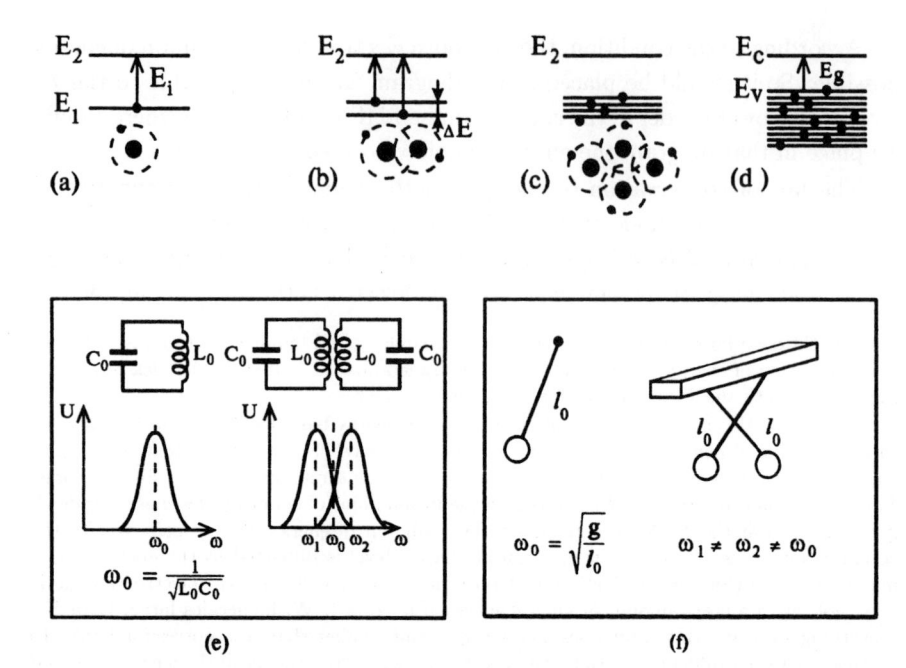

Fig. 17. Energy level splitting due to interaction: (a) A single atom. The ionization energy is E_i. (b) A system of two interacting atoms: the energy level E_1 has been split due to the interaction. (c), (d) The number of levels which appeared as a result of the splitting is equal to the number of the interacting atoms. (e) Resonance frequency splitting of two coupled LC circuits. (f) The system of two coupled identical pendulums have two resonance frequencies ω_1 and ω_2. $\omega_1 \neq \omega_2 \neq \omega_0$.

between the inductances of the circuit should not be too small). Or, speaking about it in detail, part of the magnetic flux, generated by the inductor, should penetrate into the coils of the neighbouring inductor. Then we can make sure that such a system of two circuits connected to each other has a resonance characteristic, shown in Fig. 17(e), to the right. The resonance frequency w_0 has "split" into 2 close frequencies w_1 and w_2. The stronger the coupling between the interacting circuits, the greater the difference between the frequencies $\Delta\omega = \omega_2 - \omega_1$.

Figure 17(f) shows an example of splitting the energy level into two identical pendulums l_0 long, interacting with each other. If the pendulums are isolated, the circular frequency of oscillations of both of them is equal to $\omega_0 = \sqrt{g/l_0}$, (where g is the free fall acceleration). Provided there is an interaction between the pendulums, both of them being suspended on the same beam, not very rigid, the oscillations will be characterized not by one but two frequencies ω_1 and ω_2, with $\omega_1 \neq \omega_2 \neq \omega_0$.

Now, being acquainted with the phenomenon of splitting the energy levels while they are interacting, we understand that we cannot speak of one single value of the "electron energy, bound in the interatomic orbits of silicon atoms." A semiconductor crystal presents a system of closely interacting atoms, their density being $\sim 10^{22}$ atoms/cm^3. Therefore every cubic centimetre of the crystal contains a great number of energy levels ($\sim 10^{22}$) corresponding to the valence electrons, bound in the atomic orbits [Fig. 17(d)]. The magnitude of the split, i.e. the energy gap between the neighbouring levels is very small, so that there is practically continuous spectrum of the energy of valence electrons. This region of energies is called the valence band. We can see that the E_v level, which we have mentioned before when discussing the energy diagrams, corresponds to the highest level of all possible levels of valence electrons, or, to *the top of the valence band*.

A quantum of light, whose energy is E_g, can transfer an electron from the E_v level to the conduction level E_c. And what about a quantum of light whose energy is greater? Such a quantum, provided its excess energy $\Delta E = E_{ph} - E_g$, transferred to a hole, will grasp the electron not from the E_v level, but from a much deeper level of the valence band [Fig. 18(a)]. The greater the excess energy of the quantum ΔE, the greater the energy of the newly formed hole, and the lower the hole on the energy scale. We can trace the fate of the "hot" hole emerged in that way [Fig. 18(b)]. We know the electrons tend to occupy the lowest energy state. As soon as a vacant level appears in the depth of the valence band, not filled by an electron (a hole), the electrons from higher levels will tend to take it up [Fig. 18(b)]. The vacant level (the hole) will thus tend to rise to the top of the valence band. The greater the kinetic energy of the hole Δ_E, the deeper we know it to be in the energy diagram. Thus, it is clear that while the hole moves upward in the energy diagram, it loses its kinetic energy ("it gets cooler").

Figure 18 shows that a vast region of energies below the E_v level is accessible to holes. This region is called a valence band. If a semiconductor crystal has no impurities or defects forming the energy levels lying between the levels E_c and E_v, then electrons cannot have any energy lying in the area between E_c and E_v. This area of energies is called the *forbidden band*.

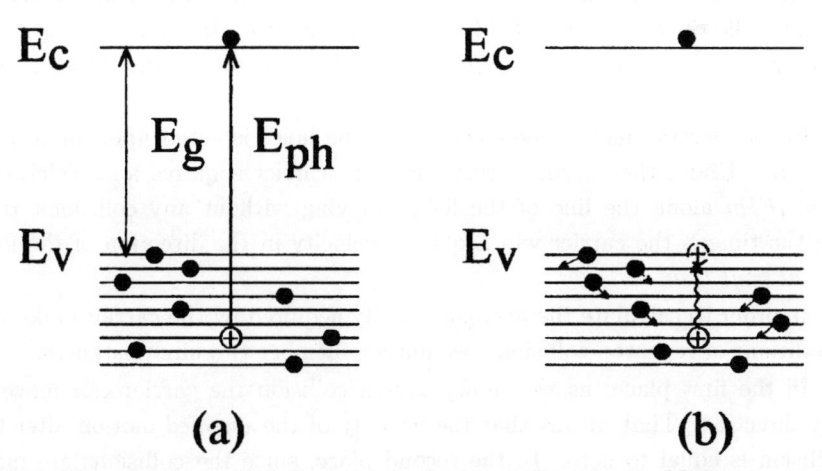

Fig. 18. The appearance of a hole in the depth of the valence band. (a) The greater the excess photon energy $\Delta E = E_{ph} - E_g$ the greater the energy of the newly-formed hole, and the lower will be the hole on the energy scale. (b) The hole loses its kinetic energy ("it gets cooler") and moves upward in the energy diagram.

So, at the temperature T the average energy of conduction electrons and holes is defined by Eq. (11). However, there exist much more "energetic" carriers whose energy is much greater than the average energy. In the energy diagram the electron's movement up the energy scale corresponds to its increase of energy, while the increase of the hole's energy corresponds to its movement down the energy scale.

2.2 Motion in the Electric Field

A drift of particles (electrons and holes) directed along the electric field is added to the chaotic thermal motion.

If there is no electric field, then the electron (or the hole) takes part only in the chaotic thermal motion. There is no directed motion whatsoever, there being no such direction which the electron (the hole) would prefer. So, although the carriers move chaotically with a very great velocity $\sim 10^5$ m/s, the velocity of the directed motion equals zero.

In the presence of the electric field, the electrons (the holes) continue their mad dance. The frequency of their collisions is, as before, $\sim 10^{12} - 10^{13}$ times per second. It is still impossible to foretell where the carrier will flow after the next collision, whether forward or backward, to the right or to the left, upward or downward, at what angle and in what direction. But no matter where it flows, the electric field will always draw it, though perhaps quite weakly, in the same direction, which will result in the appearance of a directed motion.

In the electric field F the electron (or the hole) is acted upon by a force $f = qF$. Under the action of this force, the carrier acquires an acceleration $a = qF/m$ along the line of the field. Moving without any collisions, during the time t, the carrier will acquire a velocity in the direction of the field $v = at = qFt/m$.

In order to calculate the average velocity acquired by the carrier under the conditions of repeated collisions, we must remember two circumstances.

In the first place, as we know, after a collision the carrier can move in any direction. That means that the velocity of the *directed* motion after the collision is equal to zero. In the second place, since the collisions are quite accidental, the time of the carrier's "free flight" can also be quite different. This can be clearly seen in Fig. 19, showing the velocity-time dependence of the directed motion.

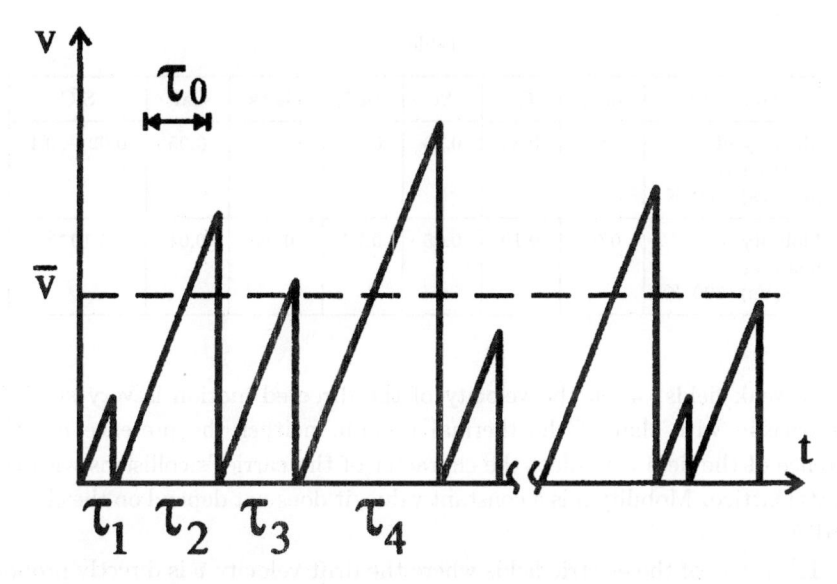

Fig. 19. The velocity of the directed motion of the electron (or hole) in the electric field versus time at random scattering.

The average velocity of the directed motion $\bar{v}$ is equal to the product of acceleration and the average time between the collisions τ_0

$$\bar{v} = \frac{q \cdot \tau_0}{m} F = \mu \cdot F \qquad (14)$$

So, the velocity of the directed motion which is often called the *drift velocity* of free carriers in a crystal is proportional to the electric field.

We've already clarified the main difference between the electron which is actually free and the so-called "free" electron in a semiconductor crystal, the conduction electron. In the crystal, the chaotic collisions being rather frequent, it is not the acceleration but the velocity of the conduction electron (or the hole) that is proportional to the field.

Coefficient of the proportionality μ in Eq. (14) is called *mobility*:

$$\mu = \frac{q \cdot \tau_0}{m} \qquad (15)$$

Table 2 gives the values of the mobility of electrons and holes of the most important semiconductor materials at room temperature.

Table 2

Semiconductors	InSb	Ge	Si	InP	GaAs	GaP	SiC
Mobility of electrons μ_n m²/(Vs), 300 K	8	0.39	0.13	0.55	1	0.05	0.02 – 0.1
Mobility of holes μ_n m²/(Vs), 300 K	0.07	0.19	0.05	0.07	0.04	0.01	5.10^{-4}

In weak fields, when the velocity of the directed motion is very small in comparison with that of the thermal motion, neither the presence nor the absence of the field can affect the character of the carrier's collisions with the crystal lattice. Mobility μ is a constant value, it does not depend on the electric field F.

The range of the electric fields where the drift velocity $\bar{v}$ is directly proportional to F is often called *linear* or *Ohmic* region. Depending on the material and temperature, the values of the fields of this region $F \leq 10^5 - 10^6 \text{V/m}$, the drift velocity being $\bar{v} \leq 10^6 \div 10^7$ cm/s. The region is linear because the velocity $\bar{v}$ grows linearly with the increase of the field F, and it is ohmic because it is this $\bar{v} - F$ dependence that corresponds to Ohm's law. Indeed, the linear dependence of the current I on the voltage U corresponds to Ohm's law. Or, which is the same, the current density j is proportional to the intensity of the field F. It is easy to see that j is proportional to F, provided the mobility μ is a constant value.

$$j = q \cdot n_0 \cdot \bar{v} = q \cdot n_0 \cdot \mu \cdot F = \sigma \cdot F \tag{15a}$$

The value $\sigma = q \cdot n_0 \cdot \mu$ is called *conductivity*, and expression (15a) is often called Ohm's law in a differential form.

2.2.1 "Hot" electrons

When the energy, acquired by the carriers under the action of the electric field, becomes equal to the energy of thermal motion, the mean time between the collisions τ_0 as a rule begins to decrease. In very strong electric fields in the majority of cases the time between the collisions τ_0 will be inversely proportional to the intensity of the field F.

In accordance with Eqs. (14) and (15), that means that the drift velocity of such "energetic", "hot" carriers will not depend on the electric field F.

The dependencies of the drift velocity on the field for Ge and Si are shown in Fig. 20. It is seen that in the weak electric fields the velocity is always proportional to the field. This part of the curve is called the Ohmic or linear region. The stronger the field, the slower the growth of the drift velocity, and when the field is strong enough, the velocity ceases to depend on the field F altogether. (This is the region of a *saturated velocity*). The smaller the mobility of the carriers in a weak electric field, the stronger the electric field must be for the velocity to be saturated. In germanium when the field is $F \geq 5$ kV/cm, the velocity of electrons in practice does not depend on the field any longer. The velocity of holes saturates for $F \geq 15$ kV/cm. The values of the saturated velocity v_s for both the electrons and holes are approximately the same, and are equal to $\sim 5 \cdot 10^6$ cm/s. In Si, the electron saturated velocity is $v_s \sim 10^7$ cm/s, and it is achieved when $F \geq 15$ kV/cm. For holes in Si, the value of the saturated velocity is also $\sim 10^7$ cm/s, but it is reached in much stronger electric fields: $F \geq 60$ kV/cm.

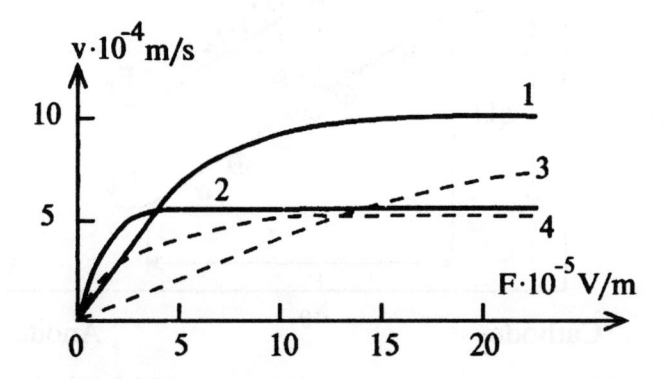

Fig. 20. The velocity of the directed motion (the drift velocity) versus the electric field in Ge and Si (T = 300 K). Curves 1,3 - Si; 2,4 - Ge. Solid lines are the electron $v(F)$ dependencies, dashed lines are the hole $v(F)$ dependencies.

2.2.2 Band diagrams

In case there is an electric field in a semiconductor, the energy diagram describing the behaviour of electrons and holes acquires a new form. To have

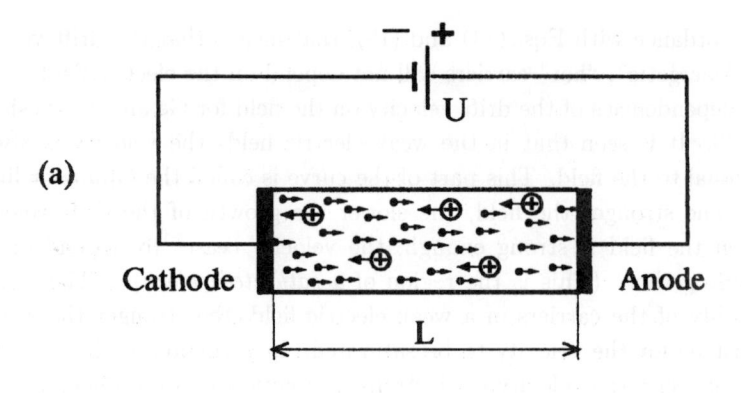

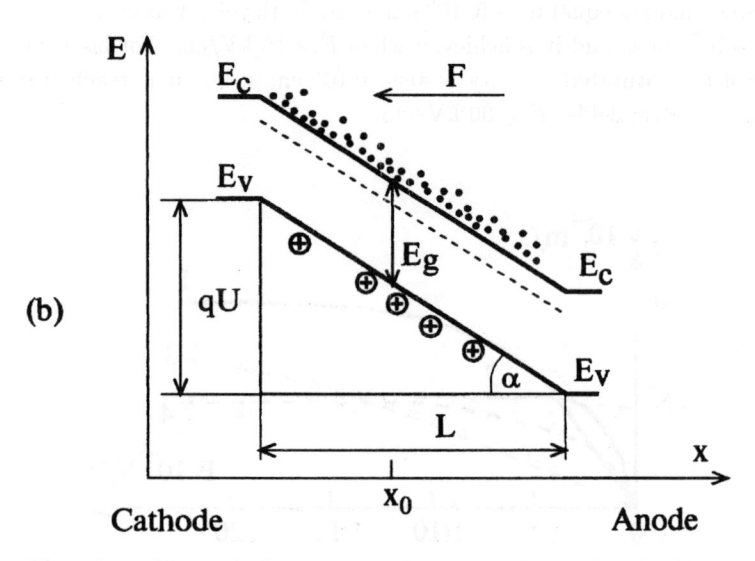

Fig. 21. The voltage U is applied to a semiconductor sample of L length. (a) the drift of electrons and holes across the sample; (b) band diagram.

a better understanding of it, let us first discuss the simplest case: a voltage U is applied to a piece of semiconductor whose length is L (Fig. 21).

There is a uniform electric field in the semiconductor $F = U/L$. Under the action of the electric field, electrons move from the cathode to the anode. The holes move in the opposite direction from the anode to the cathode. There is a potential difference U between the cathode and the anode. That means that the potential energy of the electron which is near the cathode differs from the potential energy of the electron near the anode by value $E = qU$ (q being the charge of the electron). When moving from the cathode to the anode, the electron, if it did not collide with the semiconductor lattice, would be driven away by the electric field F, as a result of which it would lose its potential energy qU which it had near the cathode, and would have exactly the same kinetic energy near the anode. But, as we know, moving in the semiconductor, the electron very often collides with the lattice. As a result of that, the electron does not acquire any additional kinetic energy while it moves towards the anode, and the whole stock of its potential energy qU is transferred to the crystal lattice of the semiconductor and is released as Joule heat.

For the positively charged hole which moves in the electric field from anode to cathode, the potential energy is higher at the anode. The hole also loses this energy on colliding with the lattice while it moves towards the cathode contact.

For the case shown in Fig. 21(a), the potential U changes linearly with the coordinate. That's why the energy diagram looks as shown in Fig. 21(b). The energy, say, of an electron which is in the conduction band at the level E_c at the cathode is larger than the energy of the similar electron at the anode by the value of qU. The same refers to the electrons which are, say, in the forbidden band at the donor level [indicated by a dotted line in Fig. 21(b)], or at any other level, deep or shallow, donor or acceptor, electrons at the E_v level, etc.

And what about the holes?

The energy of a hole, say, in the valence band at the E_v level at the cathode is *smaller* than the energy of a similar hole at the anode by the value qU. One should remember that the larger the energy of the hole, the lower is its position in the energy diagram (see Fig. 16).

The dependence of the energy on the coordinate $E(x)$ acquires the shape of a "hill". The ping-pong balls (electrons) roll down the hill, from cathode to anode. The same balls in a bucket of water (the holes) flow upward the energy scale, moving from anode to cathode.

It is very important that the parameters describing the birth and death of the electron or hole at any point of the crystal do not depend on whether there is or there is not any electric field in the sample. It can be seen that, say, at point x_0 the electron must have the same energy E_g as before in order to leave the valence band for the conduction band. The application of the electric field does not affect the ionization energy of the donor or acceptor levels ΔE or the generation and recombination velocities through the deep levels.[a]

The greater the voltage applied to the sample, the steeper is the hill. Figure 21 indicates that the tangent of the angle α between the lines of the diagram and the axis of abscissas is proportional to the intensity of the field F in the sample: $\tan \alpha \sim qU/L \sim F$. And what will the energy diagram look like in case the field in the sample is nonuniform, i.e. the intensity of the field F has different values at different points of the sample? This situation is rather common when studying semiconductor devices.

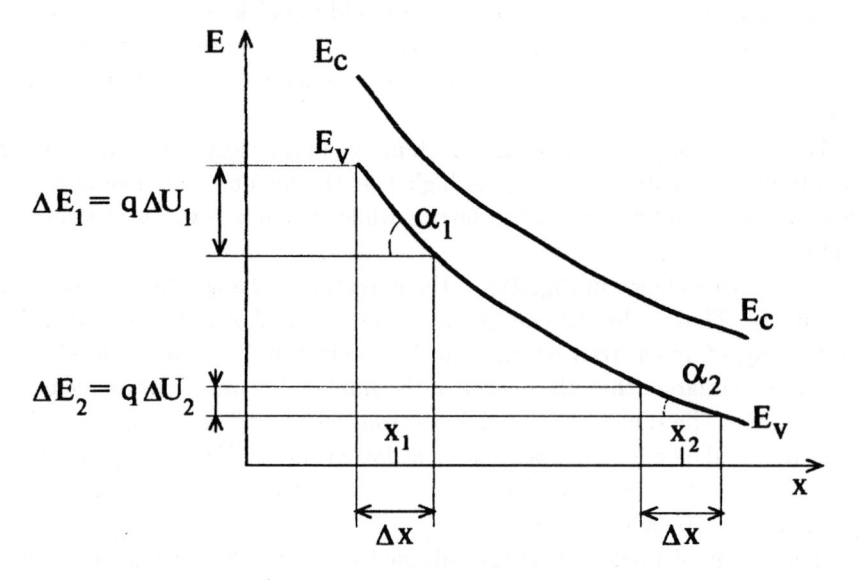

Fig. 22. Band diagram for a nonuniform field distribution.

Figure 22 shows an energy diagram for the case when the voltage in the left part of the sample near the point x_1 falls much quicker than in the right part near the point x_2. We can see that

$$\tan \alpha_1 \approx q \frac{\Delta U_1}{\Delta x} \approx F_1; \quad \tan \alpha_2 \approx q \frac{\Delta U_2}{\Delta x} \approx F_2; \tag{16}$$

---

[a]If the intensity of the external electric field F reaches very large values ($10^6 - 10^8$ V/cm), comparable to those of interatomic fields determining the conditions of birth and death of electrons and holes, then the values E_g, ΔE, the velocities of generation and recombination will greatly depend on the field F. But we are not considering these cases in this book.

It's enough to cast a look at the energy diagram to get a vivid idea of how the intensity of the field is distributed in the specimen. However, it is neither a single nor the most important advantage of the energy diagram.

As we can see in the second part of this book, the energy diagram is especially helpful when not only the electric field, but also diffusion play an important role in the motion of carriers.

2.3 Diffusion

The word diffusion comes from the Latin word *"diffusio"* (overflow, penetration). This notion can be applied to gases, liquids and solids.

The smell of the perfume spilled in a room spreads throughout the flat. Even if the radiator is switched off, the window is closed and all the chinks are caulked up to make the air absolutely motionless, the molecules of the perfume will eventually penetrate into other rooms due to the process of diffusion.

A thin coating of substance containing a lot of impurity is applied to the surface of a semiconductor plate thoroughly cleaned of any impurity. Some time later the impurity can be found in the semiconductor plate quite far from the surface. The impurity gets deep into the plate due to the process of diffusion. The rate of the impurity diffusion in a semiconductor increases very much with the increase of temperature. At room temperature it would take decades for the impurity to penetrate into the volume of a semiconductor, but when the temperature is high enough, the process of diffusion can take a few hours or even a few minutes, (Incidentally, the diffusive introduction of impurities is widely used when manufacturing semiconductor devices.)

What is there in common between the phenomena described above? It is what makes the very essence of the process of diffusion — the spontaneous penetration of the substance, unaffected by any outer influence from the place where it is in a great quantity to the place where its quantity is scarce.

The process of diffusion is a direct consequence of the chaotic thermal motion of atoms (or molecules). It is very important to understand why the disordered chaotic motion of particles leads to the directed displacement of the substance from an area of high density to an area of low density. The answer is quite simple.

Let us discuss the case with the spilled perfume. Let us imagine a hemisphere surrounding the spilled liquid by which the molecules can penetrate quite easily. We will count the perfume molecules which pass through the surface of the hemisphere from within and from outside. The perfume molecules collide with the molecules of the air and scatter in different directions. Those of them which approach the surface of our imaginary hemisphere, being kicked in a certain direction, can cross the surface. If the density of the perfume molecules were equal on both sides of the hemisphere, then the stream of molecules directed outside would be equal to the stream of molecules directed inside.

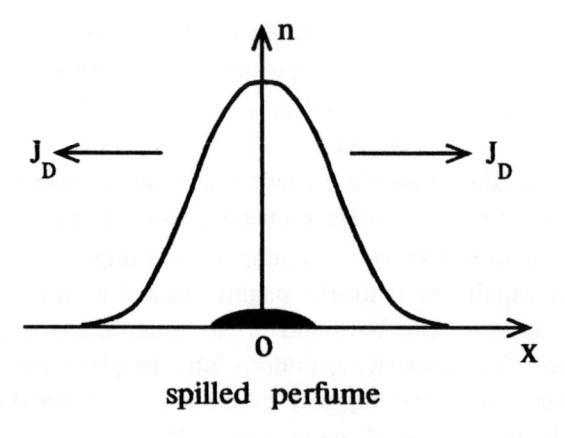

Fig. 23. The distribution of the perfume molecules over the spilled perfume. In the region $x > 0$ the derivative is $dn/dx < 0$ (n decreasing with the growth of x), the diffusion flow of carriers is $J_D > 0$; In the region $x < 0$ the derivative is $dn/dx > 0$, the diffusion flow of carriers is $J_D < 0$;

But in our case things are different. The density of the perfume molecules n is great at the place where the perfume was spilled and it becomes smaller the farther we move from it (Fig. 23). Outside the surface, the perfume density will be lower than inside it. Under the influence of the chaotic motion, the same fraction of the molecules which are close to the surface of our imagined hemisphere will cross it in both directions from outside and inside. The same fraction but of a different quantity, because the density is higher inside the hemisphere than outside it. Therefore a greater number of the perfume molecules will cross the surface from the inside than from the outside. A stream of molecules is formed, directed outside — the stream which appeared as a result of a chaotic motion. That is how the process of diffusion is formed, the process which results in equating the densities.

From what has been said it is clear that the greater the density difference on both sides of our imaginary surface, the greater will be the molecular stream caused by diffusion. The density difference determines the relation between the two streams moving in opposite directions. The rate of change of density n with the x coordinate is characterized by the derivative of the density with respect to the coordinate dn/dx. Thus the diffusive flux of carriers is expected to be proportional to dn/dx

$$J_D = -D\frac{dn}{dx}. \tag{17}$$

Why is there a minus before the right part of that equation? Figure 25 helps us to answer that question. Wherever the derivative dn/dx is positive, the flux is directed against the positive direction of the x axis and vice versa.

The proportionality factor D is called the *diffusion coefficient*.

2.3.1 Diffusion coefficient

The diffusion coefficient D in Eq. (17) is defined by the thermal velocity of molecules v_T and the mean length of the free path of molecules l. Moving with a thermal velocity v_T, the molecule can cover different distances between the collisions (see Fig. 19). However, the *mean length* of the free path ℓ certainly equals the product of thermal velocity and the mean time between the collisions τ_0:

$$\ell = v_T \cdot \tau_0 \tag{18}$$

The greater is ℓ, the more seldom the molecule changes its path after collisions with other molecules. In our example with the spilled perfume, it is clear that if the collisions with the air molecules did not impart to the motion of the perfume molecules any character of chaotic wandering, then, starting from the surface of the perfume with a velocity v_T, the perfume molecule might reach the end of the room whose length is L in time $t \sim L/v_T$. But in reality it will take more time to get there. The more frequent the collisions with air molecules, the shorter the free path ℓ, the slower the perfume molecules diffuse and the smaller the diffusion coefficient D.

With the growth of the thermal velocity of molecules v_T, it is natural to expect the diffusion coefficient to increase.

A rigorous conclusion shows that the diffusion coefficient D is connected with the values v_T and ℓ by a very simple relationship:

$$D = \frac{1}{3}\ell \cdot v_T \tag{19}$$

From Eq. (19) it is clear that the dimension of the coefficient D is $[D] = m^2/s$.

Now let us leave behind the perfume molecules and let us return to our old friends — the free carriers in a semiconductor.

There is a simple relationship between the diffusion coefficient of any particles and their mobility. We will rewrite Eq. (19) as

$$D = \frac{1}{3}v_T^2 \cdot \tau_0 \tag{20}$$

and will substitute into it the square of the thermal velocity from Eq. (11)

$$D = \frac{1}{3}v_T^2 \cdot \tau_0 = \frac{kT}{m}\tau_0 \tag{21}$$

Comparing Eqs. (21) and (14) for the mobility of the particles, we obtain

$$D = \frac{kT}{q}\mu \tag{22}$$

Knowing the value of the mobility at a given temperature, we can easily find from Eq. (22) the value of the diffusion coefficient.

The relation between the diffusion coefficient and mobility, established by Eq. (22), is valid not only for electrons and holes but also for any particles both charged and uncharged,[b] moving either in the gravity field or in the electric field. This relation is a vivid reflection of the fact that both the directed motion of particles under the action of force and the process of diffusion are hindered by one and the same process — by the collisions of particles which occur within mean time intervals τ_0 with the average thermal velocity of particles v_T.

---

[b]For the uncharged particles, mobility μ is defined as the relation of the average velocity of the particles $\bar{v}$ to force f, under the action of which this velocity is established: $\mu = \bar{v}/f$. In this case, the relationship between the diffusion coefficient and mobility is as follows: $D = k \cdot T \cdot \mu$.

The connection between the diffusion coefficient and mobility was established by Albert Einstein, and Eq. (22) is called the *Einstein relation*.

Substituting into Eq. (22) the Boltzmann constant k and the electron charge q, it is not difficult to make sure that at room temperature (300 K) the value kT/q is equal to about 0.026 V. Thus (see Table 2) at room temperature, the diffusion coefficient of the electrons in InSb is about 0.2 m^2/s of the electrons in Ge, it is 0.01 m^2/s, and in GaAs, 0.025 m^2/s.

2.3.2 Diffusion current

Let us assume that one part of a semiconductor has a greater carrier density than the neighbouring sections. This may happen if a certain section of a semiconductor is heated or illuminated. Then as we know, due to diffusion, electrons will flow from a higher to a lower density region. But the directed flow of electrons is by definition an electric current! Knowing the distribution of the carriers $n(x)$, we can easily find the density of this diffusion current.

$$j_D = qJ_D = -qD\frac{dn}{dx} \qquad (23)$$

The density of the diffusion current is equal to the flux density of the charge carriers defined by Eq. (17) multiplied by the electron charge.

When calculating the diffusion current, we should bear in mind the charge of the carriers whose behavior is being studied. If the charge is negative (the electrons), then though the flux of electrons will be directed towards the region of the lower carrier density, the electron diffusion current will flow in the opposite direction towards the higher density: $j_{Dn} = |q| \cdot D_n \cdot \frac{dn}{dx}$. In this equation, D_n is the diffusion coefficient of the electrons.

If the charge is positive (the holes), the direction of the diffusion current is the same as that of the hole flux towards the lower density: $j_{Dp} = -|q| \cdot D_p \cdot \frac{dp}{dx}$, where D_p is the diffusion coefficient of the holes.

The diffusion current is a real current no different from the current which appears under the action of the electric field F. When passing, it produces Joule heat, just like the conduction current (i.e. the current which appears under the action of the field). The diffusion current causes the deflection of the magnetic needle.

Diffusion current plays a very important role in the work of the majority of semiconductor devices.

2.3.3 Diffusion length

Suppose we are able to watch the motion of a single carrier, electron or hole. Colliding with the impurities, defects, and thermal oscillations of the lattice, the carrier rushes chaotically, changing after every collision the direction of its motion (Fig. 24). Between collisions, the carrier (for the sake of clarity in the future, we will speak of the electron) covers the distance equal to the mean free path ℓ. At what distance (L) from the starting point will the electron be after N collisions?

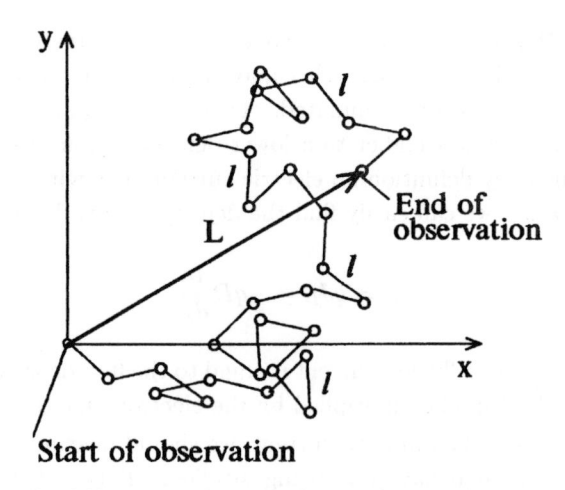

Fig. 24. As a result of N steps made at random, each step l long, the object covers the distance $L \approx l \cdot \sqrt{N}$.

This problem of random wanderings is known in mathematical statistics as "the path of the drunk." For both the drunk and the electron (though for different reasons) the direction of every next step is quite unpredictable and does not depend at all on the previous step. Therefore the problem we are interested in can be formulated in the following way: How far beyond the starting point will the drunk man have gone after he has made N steps, each step ℓ long?

It goes without saying that if we did not speak of the drunk man, the answer would be quite evident: having made N steps, each step ℓ long, the man would be at the distance $L = N\ell$ from the starting point. But from Fig. 24 and from our own observational experience, we know that the path covered by the drunk man after he had made N steps will be shorter.

A very simple though somewhat cumbersome conclusion (it can be found in our book "Getting to know Semiconductors, World Sci, 1992) shows that for the drunk man (and for the electron) after N steps, each step ℓ long, the distance covered will be

$$L = \ell \cdot \sqrt{N} \qquad (23)$$

Let us rewrite Eq. (23) in another form. Let the electron move in a certain direction under the action of diffusion during the time t. During that time, on the average, every electron will make $N = t/\tau_0$ "steps" ℓ long, where τ_0 is the mean time between the collisions, l is the length of the free path. Thus, *on the average*, the path covered under the action of diffusion during time t will be equal to [see Eq. (18)]:

$$L \cong \ell \cdot \sqrt{\frac{t}{\tau_0}} \cong v_T \cdot \tau_0 \sqrt{\frac{t}{\tau_0}} \cong \sqrt{v_T^2 \tau_0 \cdot t} \cong \sqrt{D \cdot t} \qquad (24)$$

Equation (24) establishes a most unusual distance-time dependence. In accordance with Eq. (24), the time necessary to cover the distance L on account of diffusion is proportional to L^2:

$$t \cong \frac{L^2}{D} \qquad (25)$$

The sizes of modern semiconductor devices are often of the order of a micrometre (10^{-6} m) or even a fraction of a micrometre. Assuming that the value $D = 10^{-2}$ m^2/s (see the previous section), we conclude that the electron covers on account of diffusion the path $L \sim 10^{-6}$ m during the time 10^{-10} s and the path $L \sim 10^{-7}$ m during 10^{-12} s! Thus, in modern devices the diffusion transport is a rapid and important process.

In some types of devices, a certain type of diffusion, the diffusion of excess (nonequilibrium) carriers, plays the most important role. The peculiarity of this type of diffusion is that the nonequilibrium carriers, while diffusing, at the same time move away from the place of their introduction into the semiconductor, and perish (recombine).

While studying in Sec. **1.3.3 (Recombination through deep levels)** the process of recombination, we established the following: if by illuminating a semiconductor, we create excess carriers, then after the light has been switched off, the concentration of the excess carriers will diminish. Their lifetime τ, the time it takes the excess carriers to recombine, depends on the nature of

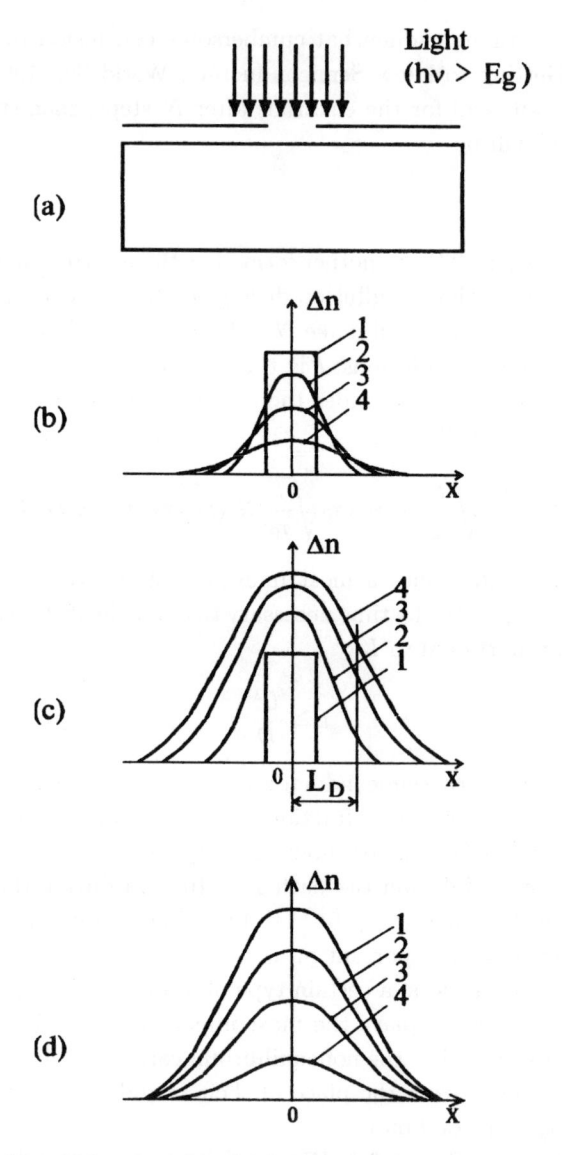

Fig. 25. The carriers which appeared in the illuminated part of the semiconductor propagate to the shaded region due to diffusion: (a) the specimen is covered by a metal plate with a slit in it; (b) the distribution of the excess carriers after a short flash of light, ($t_1 < t_2 < t_3 < t_4$); (c) the excess carrier distribution at the moments of time $t_1 \div t_4$ ($t_1 < t_2 < t_3 < t_4$) with the light switched on for a very long time; (d) the excess carrier distribution with the light switched off ($t_1 < t_2 < t_3 < t_4$).

the semiconductor, on the type and the density of the deep impurities in the material, and lies within the limits from $\sim 10^{-10}$ up to 10^{-2} s.

Let us discuss the experiment depicted in Fig. 25. The semiconductor sample is illuminated by light with quanta of energy $h\nu$, which is greater than the energy of forming an electron-hole pair E_g. The sample's surface is covered by a metal plate with a slit cut in it. The light does not penetrate through the metal, so the electron-hole pairs created by light appear only directly under the slot [Fig. 25(a)].

We switch on the light and then switch it off at once [Fig. 25(b)]. Let us assume that the flash is so short that the electrons which appeared under the action of light did not have time to either spread under the influence of diffusion or recombine. In this case, the distribution of electrons immediately after the end of the light pulse will correspond to curve 1. What will happen to the nonequilibrium excess electrons next?

As we know, the nonequilibrium carriers will recombine (die out) with a characteristic lifetime τ. Besides, they will travel under the action of diffusion from the region of high density to the region of low density. Curves 2–4 [Fig. 25(b)] indicate what happens to the pack of the excess carriers due to the joint action of the process of both, recombination and diffusion.

Electrons, created by light, recombine during the time of the order of τ. According to Eq. (24) while dying they have time to advance from the illuminated region where they have appeared to the shaded region covering the distance of the order of the *diffusion length* L_D

$$L_D = \sqrt{D\tau} \qquad (26)$$

Let us now conduct the following experiment. We will switch on the light and leave it on for an indefinite peried of time [Fig. 25(c)]. Immediately after the light was switched on, the nonequilibrium electrons existed only in the illuminated region (curve 1). Then the electron packet will begin to spread due to diffusion, though the process of generating carriers by light will dominate over the process of recombination. The concentration in the illuminated region will still increase (curves 2 and 3). Curve 4 corresponds to the stationary state. Electron distribution corresponding to curve 4 will remain stable while the light is switched on. The number of carriers created by light is exactly the same as the number of carriers perishing due to the recombination. At every point the number of carriers, delivered by diffusion from the neighbouring region with a higher concentration, is equal to the sum of the number of carriers leaving this point for the neighbouring region with a lower concentration and the number of the recombining carriers.

The nonequilibrium carriers, created in the illuminated region, penetrate into the shaded region as far as several *diffusion lengths* L_D.

Figure 25(d) indicates what will happen if the light is switched off. Nonequilibrium carriers recombine, the recombination process taking time τ.

2.4 Summary

Free carriers are in a state of random chaotic motion, which is very rapid. Under usual conditions the average velocity of such chaotic motion is $\sim 10^5$ m/s. The electric field causes a directed drift of carriers which is added to this chaotic motion. If the field is weak, the mean velocity of the drift is proportional to the electric field (Ohm's law). In a strong field the drift velocity of carriers is saturated. If the free carrier concentration in the semiconductor is non-uniform, then there appears diffusion, a flux of carriers directed from a higher density region to those regions where the density is lower.

Part II

Barriers & Junctions

The first part of this book gave us much information about those properties of semiconductors which are called *volume* properties. That knowledge is quite essential in order to understand the work of any semiconductor device.

However, that knowledge is insufficient for understanding the principles of work of diodes and transistors. In the work of diodes and transistors, a major role is played by phenomena called boundary effects. The most "natural" — inevitable for any device, any semiconductor structure — boundary is the border between the semiconductor and its environment (air or vacuum). To make our acquaintance of this new range of problems, we will begin by studying that boundary. The knowledge of the properties of that boundary will greatly facilitate the understanding of any other boundary phenomena, to say nothing of the fact that those properties are most interesting by themselves as such.

Chapter 3

The Barrier on the Crystal Boundary

> I am in prison, behind the bars...
>
> Alexander Pushkin
> 1799 - 1837

Let us assume that there is a piece of semiconductor or metal in front of us, and we know that there is a certain number of free electrons (conduction electrons) in it. If it is a metal, then every cubic centimetre of it will contain $\sim 10^{22}$ free electrons. On the other hand, there are practically no free electrons in the surrounding medium (the air) [Fig. 26(a)]. Being aware of the phenomenon of diffusion, we understand that according to every canon and every rule, there must appear a diffusion flow of electrons from the metal where they are numerous to the air which has practically no electrons at all. And just as the pool of split perfume evaporates, the free electrons from the metal or the semiconductor should "evaporate" some time later.

Moreover, we could easily calculate the time in which the electrons should have left the crystal. By the order of magnitude, that time is equal to the time of the electron diffusion through the crystal $t_D \approx L^2/D$, where L is the size of the crystal and D is the diffusion coefficient. Handbooks usually contain the values of electron mobilities in different materials. Say, for gold (Au), the mobility is $\mu = 30$ cm^2/(V.s). Using the Einstein relation (22), we find that the diffusion coefficient in Au is about 0.75 cm^2/s. Thus, from a piece of gold whose size is 1 cm, the conduction electrons had to disappear on account of diffusion during the period of time $t = 1.3$ s.

But, in reality, nothing of this kind happens at all. And the gold decorations of Troya which are about three thousand years, and even the native ore of gold,

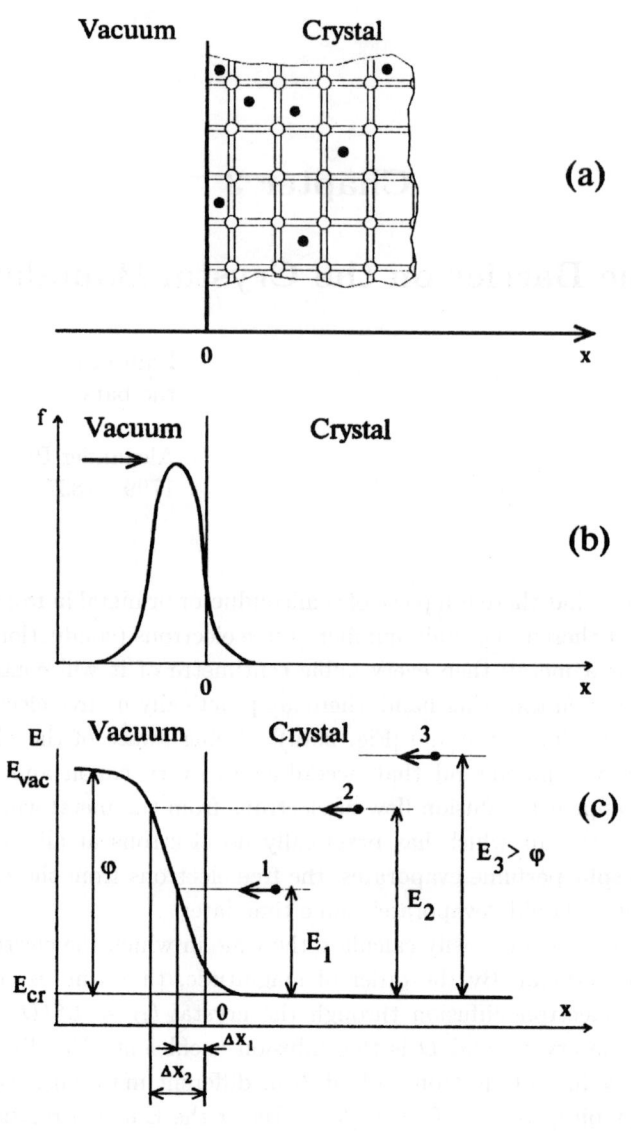

Fig. 26. The barrier on the crystal boundary. (a) Conduction electrons are blocked in the crystal and cannot leave it. (b) There is a force f in the vicinity of the boundary which prevents electrons from leaving the crystal. (c) In order to leave the crystal the electron must overcome the potential barrier of a height $\varphi = E_{vac} - E_{cr}$. The electron with kinetic energy E_1 penetrates into the barrier and cover the distance Δx_1. Then the electron will turn backwards. The electron with kinetic energy $E_3 > \varphi$ can leave the crystal.

which had been lying in the air millions of years contain the same concentration of conduction electrons as a bar of gold just out of the smelting furnace.

3.1 Work Function

What is it that prevents electrons from leaving the metal? Thinking about it we come to the conclusion that at the demarcation line between the crystal and the surrounding medium (for the sake of simplicity we will for the time being consider it to be a vacuum) the electron must be affected by a certain force preventing it from leaving the crystal. This force acts upon electrons in the vicinity of the border and it disappears some distance from the border both in the depth of the crystal and outside it [Fig. 26(b)].

It is clear that the force holding the charged particles — electrons — is the electric field. Thus Fig. 26(b) gives a qualitative picture of the distribution of the electric field near the surface of the crystal.

The same idea of force preventing the electrons from leaving the crystal can be clearly demonstrated in the language of band diagrams [Fig. 26(c)].

Let us imagine that an electron in the crystal approaches the border. The velocity of the electron is directed perpendicular to the border. The electron tends to leave the crystal. But as soon as it gets into the area affected by the force $f = q \cdot F$ holding it within the crystal [Fig. 26(b)], its velocity begins to decrease. For if the particle is moving in the region affected by the force f, directed against the motion, in order to overcome that force and cover the distance Δx, the particle must perform the following work $\Delta A = f \cdot \Delta x = qF \cdot \Delta x$. The kinetic energy of the particle will decrease by the value ΔA. If the kinetic energy E of the particle which has flown up to the border is not large, the particle will spend this energy very soon, moving within the field of the braking force $f = q \cdot F$. Its velocity will drop to zero. Then the electron will turn backwards, and will accelerate under the action of the same force, and will flow right into the crystal.

The greater the kinetic energy of the electron flying up to the border, the farther it will penetrate into the area where the braking force f is acting. Finally, we can imagine that there will be a very energetic electron which will manage to go all the length of the region where the braking force f is acting and fly out of it. The minimal kinetic energy the electron must have to perform it is called *work function* and is usually designated by the Greek letter φ [Fig. 26(c)]. This work is certainly equal to the difference between the energy of the electron at rest in the vacuum E_{vac} and that of the electron at rest in the crystal E_{cr}.

So, in order to leave the crystal, the electron must overcome the potential barrier of a height φ.

How does the force which holds the electron within the crystal appear? Or, in other words, what is the nature of work function?

3.1.1 Come back, return, I call you back!

The first attempts to explain the physical nature of the work function were made at the beginning of the 20th century by two outstanding scientists: the German scientist Walter Shottky and the American — Irwing Langmuir. Their idea was quite simple. As soon as the electron leaves the crystal, the latter which up to that moment was electrically neutral goes positive. And, quite naturally, it begins to attract the negatively charged electron, and tends to attract it back.

In order to overcome that attraction and fly away, the electron which has left the crystal must possess a certain kinetic energy. Its minimal value will evidently be equal to the work done by the electron to leave the crystal. Such a mechanism of forming the electric barrier on the border must evidently exist in every material.

Langmuir and Schottky calculated the expected value of the work function for metals. It was found that in order to overcome the force of attraction and leave the crystal, the electron will have to spend the energy of $1 - 2$ eV.

However, the mechanism, considered by Langmuir and Schottky, proved not to be the only cause preventing the electron from leaving the crystal.

3.1.2 Double charged (dipole) layer

Conduction electrons keep on attacking the border between the solid body and the vacuum, striving to fly outside. The kinetic energy of the majority of electrons is certainly not large enough to overcome the barrier at the border and leave the crystal. But due to those attacks, there exists, at every given moment, a negatively charged electric cloud beyond the crystal, while inside the crystal there is a non-compensated positive charge (Fig. 27). There occurs a double charged (dipole) layer at the border, and the electron which flies into this layer is acted upon by a force $f = q \cdot F$, which tends not to let the electron leave the crystal.

Studying the barriers we will often deal with double-charged layers. So we will now discuss in more detail the effect of this layer upon a charged particle (Fig. 28). On the whole, the dipole layer is electrically neutral. The negative charge in it is exactly equal

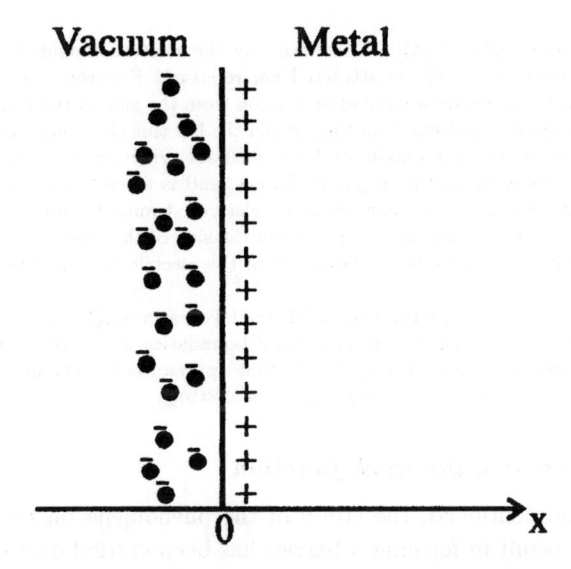

Fig. 27. The dipole layer at the metal-vacuum boundary.

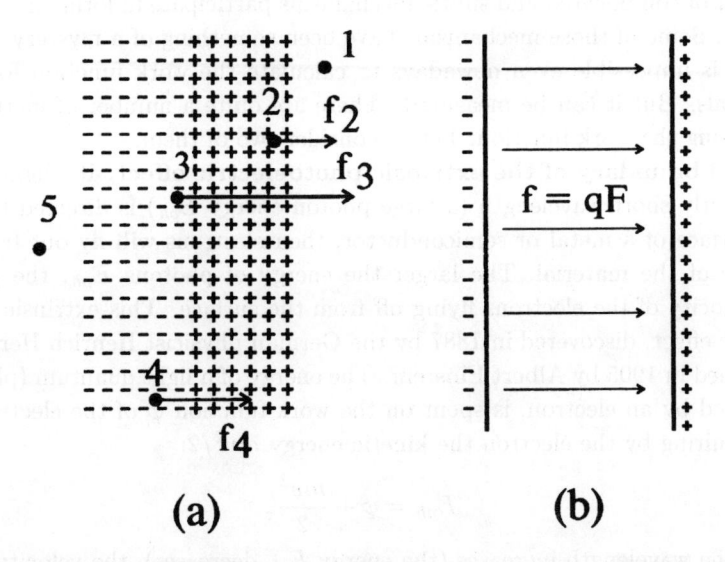

(a) (b)

Fig. 28. On the whole, the dipole layer is electrically neutral. (a) The electrons outside the charged layer (electrons 1 and 5) are neither attracted nor repulsed. The maximum force acts on electron 3. (b) As a rough approximation the action of the dipole layer is equivalent to the action of a capacitor.

to the positive charge [Fig. 28(a)]. Therefore, say the electron outside the charged layer (in Fig. 28(a) electron 1) is neither attracted nor repulsed. Electron 2, which got into the charged layer is acted upon by a force of attraction from the side of the right-hand positive charge and by a force of repulsion from the left (on the left this electron has negative charges more than those positive). The maximum force must act evidently on electron 3. The latter is repulsed to the right by all the negative charges, and is attracted in the same direction by all the positive charges. The force acting on electron 4 must be smaller, and electron 5 which is beyond the layer is not acted upon from the side of the layer by any force at all.

It is clear that the qualitative distribution of the electric field will be the same as in Fig. 26(b).

Sometimes, ignoring the distribution inside the layer, as a rough approximation, we may consider that there is a region in the vicinity of boundaries where the charged particle is acted upon by a constant force $f = q \cdot F$. In that approximation the action of the dipole layer is equivalent to the action of a capacitor [Fig. 28(b)].

3.1.3 How to define the work function

As was already mentioned, the study of the phenomena on the boundary of crystals which result in forming a barrier has been carried on since the beginning of the 20th century, i.e. for almost one hundred years. And it is still going on. The thorough investigation of properties of the surface have shown that a number of complicated and subtle mechanisms participate in forming a surface barrier. Some of those mechanisms have been something of a mystery. Therefore it is impossible even nowadays to calculate the work function for some materials. But it can be measured. There are quite a number of methods of measuring the work function. Let us consider two of them.

Red boundary of the extrinsic photoelectric effect. If a light with a sufficiently short wavelength (a large photon energy E_{ph}) is directed towards the surface of a metal or semiconductor, then electrons will fly out from the surface of the material. The larger the energy of photons E_{ph}, the greater the velocity of the electrons flying off from the surface. This extrinsic photoelectric effect, discovered in 1887 by the German physicist Henrich Hertz, was explained in 1905 by Albert Einstein. The energy of a light quantum (photon), absorbed by an electron, is spent on the work function φ of the electron and on acquiring by the electron the kinetic energy $mv^2/2$:

$$E_{ph} = \varphi + \frac{mv^2}{2}.$$

If the wavelength increases (the energy E_{ph} decreases), the velocity of the electrons flying out will decrease. And, finally, with a certain critical energy of photons E_{ph}, electrons do not fly off at all. The extrinsic photoeffect disappears. This energy is evidently equal to the work function φ.

Thermoionic emission. The essence of the thermoionic emission is indicated in Fig. 29, which is similar to Fig. 26(c). It shows again that the overwhelming majority of electrons have the energy E_{cr} inside the crystal and at the boundary of the crystal they meet with an energy barrier of height $\varphi = E_{vac} - E_{cr}$. Electrons in metal are in the energy "well" of height φ.

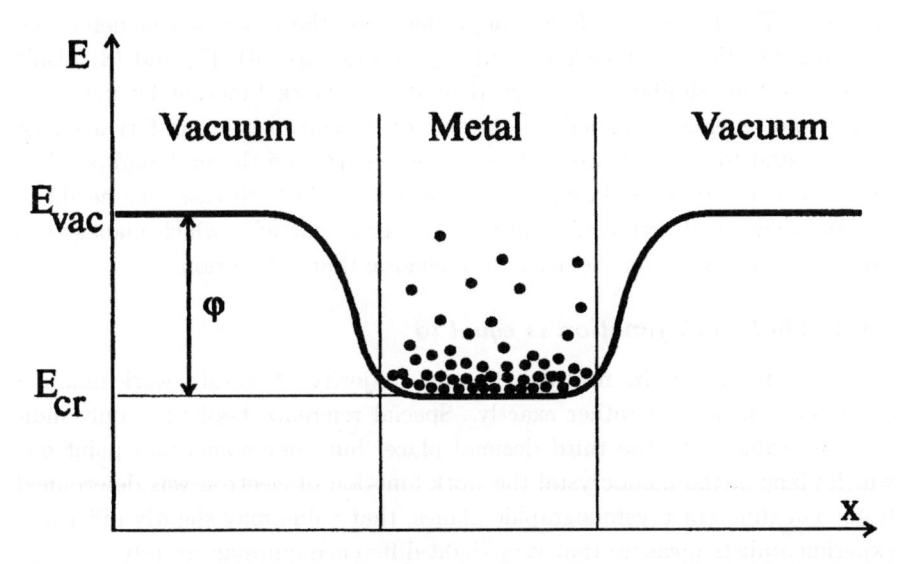

Fig. 29. The overwhelming majority of electrons have the energy E_{cr} inside the crystal. These electrons are on the bottom of the "potential wall" of φ depth. At a certain temperature T there always exists a certain number of very energetic electrons, whose energy is much greater that E_{cr}.

We know, however, that at the certain temperature T there always exists a certain number of very energetic electrons, whose energy is much greater than that of E_{cr} (see Eq. (12) and Fig. 14).

Electrons whose kinetic energy $(E - E_{cr})$ exceeds the work function φ, can leave the crystal and fly outside. The concentration of such "superenergetic" electrons can be calculated by the rough equation $n \approx n_0 \cdot \exp(-\varphi/kT)$.

The electrons which fly out of the specimen are registered the following way. The metallic or semiconductor specimen is placed into a vacuum chamber. This is done so that the motion of the electrons which flew out of the specimen might not be affected by the collisions with the molecules of the air. The specimen is applied by a "minus" and a metal plate is placed nearby, which is applied by a "plus". Under the effect of the electric field created by the battery, the electrons which flew out of the specimen (the cathode) fly towards the plate (the anode). Current is generated between the electrodes, and the current is evidently proportional to the number of electrons, which flew out of the specimen: $I \approx \exp(-\varphi/kT)$.

Then the current-temperature dependence is plotted as a dependence $\ln I$ against $1/T$. The slope of that dependence to the abscissa will determine the work function φ. Please look at Fig. 8 and Eqs. (6), (7) and (8). Isn't there a certain similarity between defining the work function by means of thermoionic emission and defining the energy of electron-hole pair formation by the temperature dependence of the concentration? And the profound physical reason, causing that similarity, is also quite clear: in both cases the height of energy barrier is determined from the number of particles which managed to overcome the barrier on the account of chaotic thermal motion.

3.1.4 What work function is equal to

Work function in metals. For the majority of metals, work function φ has been measured rather exactly. Special reference books not only indicate the value φ to the third decimal place, but they sometimes point out which plane of the monocrystal the work function of electron was determined from. For different crystallographic planes, that value may slightly differ and experimentalists measure that very slight difference quite accurately.

Work function in metals makes several electron-volts and is in the range ~ 2.1 eV (for lithium, potassium and cesium) to ~ 5.4 eV (for platinum). In tungsten, which we know very well, (filaments of incandescent lamps being made of tungsten), the value φ for different planes makes 4.3–5.3 eV. For iron, $\varphi \approx 4.7$ eV, for gold, it is ≈ 5.3 eV.

Work function in semiconductors. Figure 30 shows a band diagram which is well-known to us. As always, the bottom of the conduction band is designated by E_c, the top of the valence band by E_v. The levels of shallow donors and acceptors E_d and E_a and the deep level E_t are also shown.

Let the energy of an electron in a vacuum be equal to E_{vac}. What about the energy of an electron in a semiconductor crystal?

What is it, generally speaking, equal to?

With metals, everything was clear. All the electrons there were always free. And what about a semiconductor? As we know, in semiconductors, situations may be quite different. In the n-type semiconductor, when the temperature

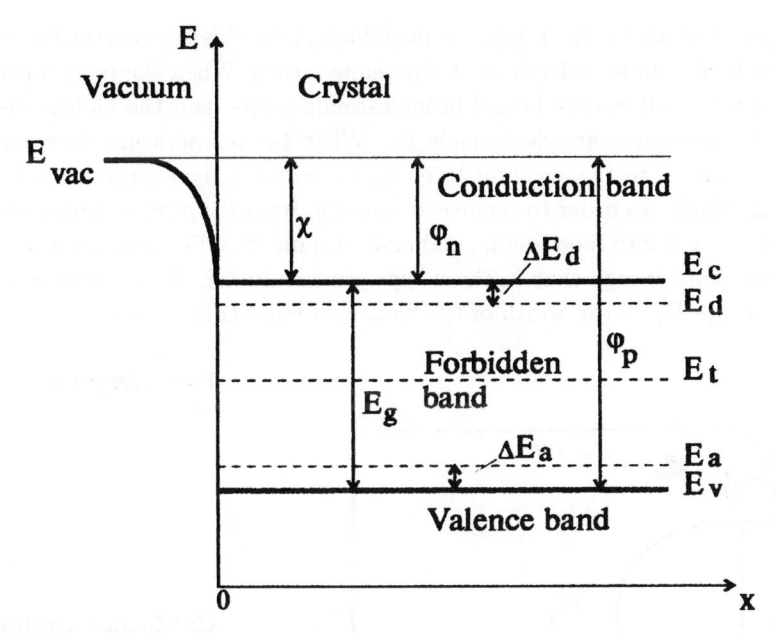

Fig. 30. The work function in the same semiconductor depends greatly on doping and temperature.

is high enough, there are many conduction electrons, and the overwhelming majority of them have the energy $\approx E_c$ [see Fig. 14 and Eq. (12)]. It would appear reasonable to expect that, as much as it is in metals, the work function in semiconductors will be $E_{vac} - E_c$ (Fig. 30).

If the n-type crystal is cooled, the concentration of electrons in the conduction band will decrease: a greater number of electrons will be bound at donor centres. With $T = 0$, as we know, there are no conduction electrons at all. All atoms of donor impurity are neutral and keep "extra" electrons to themselves. So, the highest energy level at which there are electrons in this case is the donor level E_d. To snatch the electron from the crystal, it is necessary now to spend the energy $E_{vac} - E_d$, i.e. larger than in the previous case by the value of the ionization energy of the donor level E_d. If the donor level is shallow, and the value ΔE_d is small, then the difference will not be very appreciable. But the donor centre may also be deep. Its ionization energy may be equal to decimal fractions of electronvolt. In this case the work function of the semiconductor when the temperature is low, may be much larger than at a high temperature.

And what about the p-type semiconductor? In this semiconductor, there are practically no free electrons at any temperature. When the temperature is low, all the electrons are bound in interatomic orbits, and the highest energy level at which there are electrons is E_v. When the temperature rises, part of electrons moves to the acceptor level E_a, overcoming the energy barrier ΔE_a [see Fig. 10(c)]. In order to remove an electron from the p-type semiconductor crystal, place it into the vacuum, and make it quite free (Fig. 31), it is necessary to spend more energy than in the n-type semiconductor. This energy is larger by the value E_g — the width of the forbidden band (Figs. 30, 31).

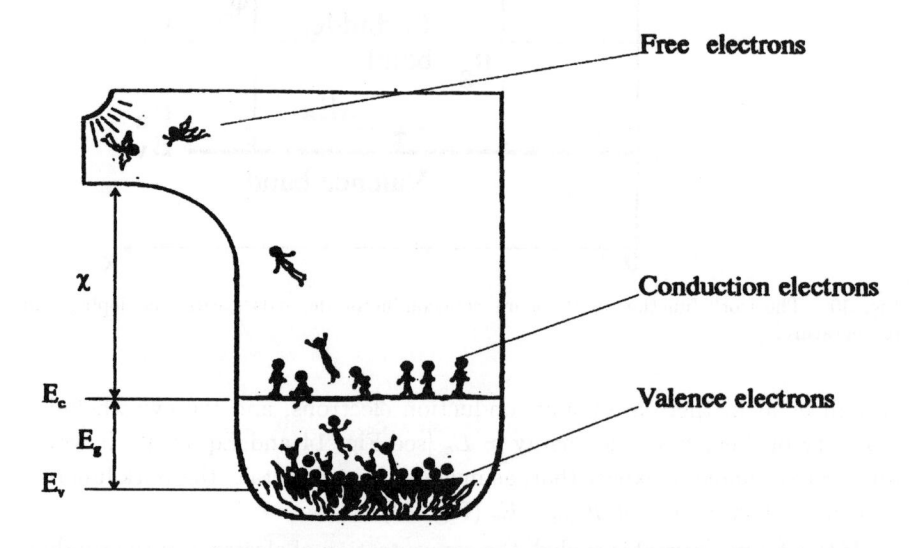

Fig. 31. Compare this figure with Fig. 5. It is necessary to spend the energy E_g to create a conduction electron. To remove the conduction electron from the crystal, it is necessary to spend the energy χ. In order to remove the electron with the energy E_v (from the top of the valence band), it is necessary to spend the energy $E_g + \chi$.

Since the work function in the same semiconductor greatly depends on doping and temperature, in order to compare the value of work function in different semiconductors, a special unit χ is introduced. This unit is called *electron affinity* of the semiconductor. As is seen from Fig. 30, $\chi = E_{vac} - E_c$ and is constant for a given semiconductor. Roughly speaking, the value χ approximates the work function φ_n of the n-type semiconductor. The value of the work function of the same semiconductor of p-type is, roughly speaking, larger by the value E_g (Fig. 30).

Discussing the magnitude of the semiconductor work function we do very well without rough estimation. If we know the concentration and the ionization energy of impurity centres introduced into the semiconductor, we can calculate quite exactly the value of work function for any temperature. Such calculations were done repeatedly, specifically for germanium and silicon, and were time and again compared with the experimental results. In many cases, the data of the calculation and those of the experiment coincided absolutely, but always for indirect experiments, i.e. for those of which it is not the work function which was measured, but some other magnitude associated with it.

In 1962, a most refined and clever experiment was made in which the dependence of work function on the level and type of doping was defined in germanium and silicon in a direct manner. Long semiconductor crystals were grown whose impurity concentration and kind of doping were changed from one point to another.

Knowing the change of the impurity concentration with regard to the coordinate, one can calculate the change of work function in the crystal according to the coordinate. Results of these calculations are shown in Fig. 32 by a dashed line.

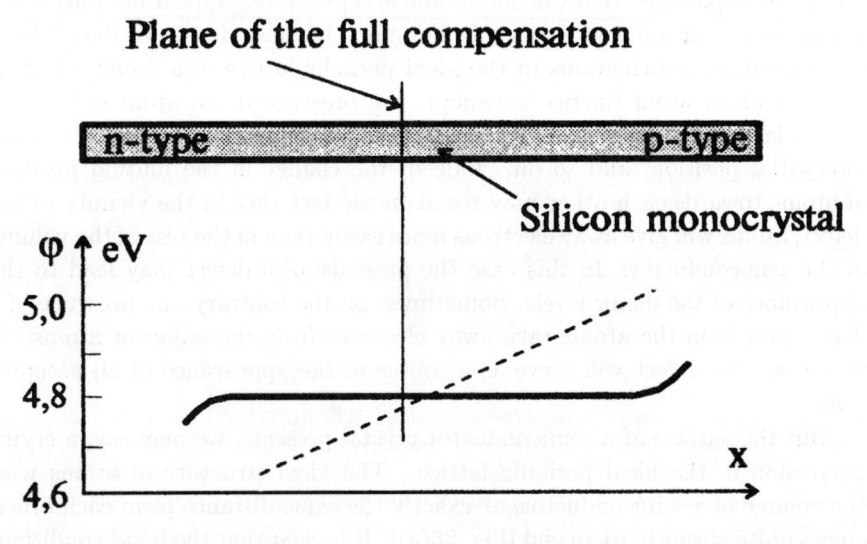

Fig. 32. The dependence of work function on the coordinate x for the nonuniform doped Si crystal. The dashed line represents the calculated dependence. The solid line represents the experimental data (experiment by Allen and Gobeli, 1962).

Then, splitting off layer after layer of the monocrystal, researchers measured the change of the work function. The result of the experiment, as we see, is quite different from what was expected. Though the precision of measurement is several hundredth and the expected change was to be a few decimal fractions of electron-volt, the measured work function did not depend on the coordinate. In any plane of the crystal, both n- and p-type, it approximated the value expected for the intrinsic silicon.

What does this wonderful result imply?

3.2 Surface States

To tell the truth, Allen and Gobeli's results did not make any sensation in 1962. They just vividly demonstrated the fact that complex phenomena on the border line between the semiconductor and environment are not confined to just forming on the boundary an energy barrier.

The second important factor defining the properties of a semiconductor in the vicinity of the border is the *surface traps*, or, as they are often called, *surface states*.

We are already quite familiar with energy levels in the forbidden band. We know about the donor and acceptor levels, shallow and deep levels. It should be emphasized that the donor- and acceptor levels appear not only when impurities are introduced into semiconductors, but also due to various defects and structural deformations in the ideal periodic lattice of a semiconductor: the lack of an atom (lattice vacancy), the presence of an atom not in the crystal lattice node of a semiconductor where it is supposed to be, but in the interstitial position, and so on. Indeed, the change in the mutual position of atoms towards each other may result in the fact that in the vicinity of the defect, atoms will give away electrons more easily than in the rest of the volume of the semiconductor. In this case the presence of a defect may lead to the appearance of the donor levels. Sometimes, on the contrary, the presence of a defect may help the atoms take away electrons from the adjacent atoms. In that case, the defect will serve as a source of the appearance of an acceptor level.

But the surface of a semiconductor crystal presents, we may say, a crying perversion in the ideal periodic lattice. The ideal structure of atoms with the volume of a semiconductor, at exactly the same distance from each other, comes quite abruptly to an end [Fig. 26(a)]. It is clear that the bond conditions between the valence electrons, which belong to atoms near the surface, will differ from those within the volume.

In 1932, Igor Tamm, the future Nobel Prize winner in physics, proved theoretically that for the appearance of additional energy levels it is sufficient to have just one break in the crystal lattice, even without the presence of any foreign atoms on the surface of the semiconductor. These surface states are called the *Tamm levels*.

A little later, in 1939, William Shockley, another future Nobel Prize winner, came to a similar conclusion, though on the basis of quite different grounds. Shockley paid attention to the fact that atoms near the surface due to the break in the lattice, have non-paired valence electrons [see Fig. 26(a)]. These conditions may facilitate the detachment of the non-paired electrons from the atoms near the surface. In this case, donor surface states are generated. And an opposite situation may also occur: non-paired electrons near the surface will acquire an ability to bind (capture) the conduction electrons which are in the volume of crystal. In this case, acceptor surface states will be generated.

The Tamm levels exist on the so-called "atomically pure" surface, i.e. the surface perfectly smooth, without any foreign atoms or structural defects.

Under ordinary conditions, i.e. in the air, the surface of the semiconductor is always covered by a layer of oxides. Besides, the foreign atoms are absorbed on the surface. Finally, there are always structural defects on the surface, such as lattice distortion, which appears in the process of splitting off or cutting the crystal, vacancies, i.e. vacant places in the nodes of the lattice, semiconductor atoms which have shifted from their usual places in the lattice nodes into interstitial positions. All those impurity and structural violations also prove to be the sources of surface states.

The assemblage, or, as it is often called *the spectrum of surface states*, of the real surface can be changed by alternating the conditions of oxidation *or* method of treating the surface, preventing the absorption of atoms; or; on the contrary, saturating the surface with atoms of a certain substance. Just like the volume levels, the surface traps can be donor or acceptor levels, shallow or deep levels. The density of surface states, measured in m^{-2} or cm^{-2}, can be used as a quantitative measure of surface state density.

As we know, 1 cm^3 of any solid contains $N \approx 10^{22}$ atoms. That means that along the edge of a cube 1 cm long, there are $N^{1/3}$, and on the side 1×1 cm there are $N^{2/3} \approx 10^{15}$ cm^{-2} atoms of the basic material. So, if for every atom of the surface of a semiconductor, there is an impurity level, then that situation corresponds to the surface state density $\sim 10^{15}$ cm^{-2}. The surface state density on actual surfaces is in the range from $\sim 10^9$ to $\sim 10^{15}$ cm^{-2}.

3.3 Bending Bands, Surface Potential

Depending on whether the surface states are mainly those of donors or acceptors, the surface will either release electrons to the volume of a semiconductor and get a positive charge, or, vice versa, will capture electrons from the volume and get a negative charge. Let us assume that the acceptor states are predominant on the n-type semiconductor surface. Then the surface levels will capture electrons from the adjacent part of the semiconductor volume. As a result of it, the surface will be charged negatively. And near the semiconductor, there will be a layer depleted of electrons, i.e. charged positively. Thus, close to the surface, a dipole layer will be formed, whose field will repulse electrons and they will go deep into the crystal. (This situation resembles the case, shown in Fig. 27. But the negative charge is now formed not on the account of electrons flying out of the crystal, but on the account of electrons captured on the surface states.) The dipole layer field is maximum near the surface and decreases as it recedes from the surface into the crystal (Fig. 33).

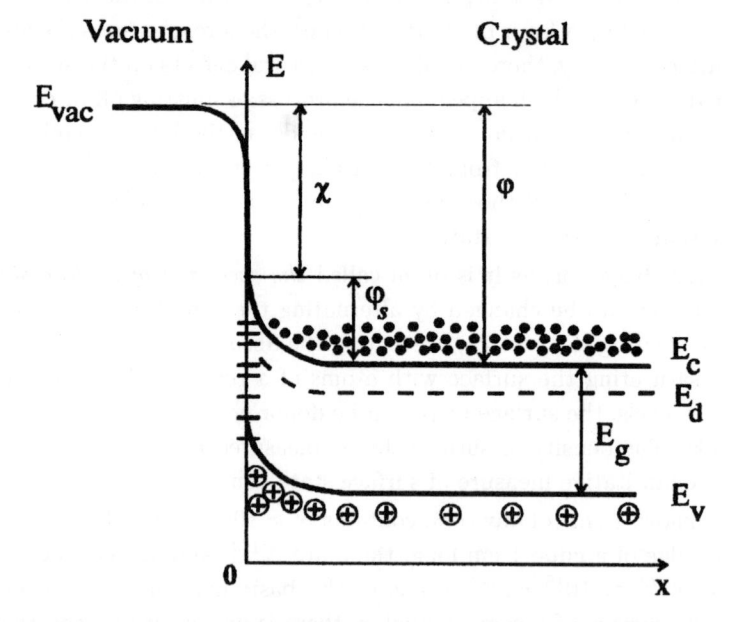

Fig. 33. Band diagram of an n-type semiconductor with negative surface states. Short lines at the crystal-vacuum interface correspond to the energy position of the surface states in the forbidden band. E_d is the volume donor level.

It would be useful to compare Fig. 33 with Fig. 30. In Fig. 30 the boundary of the band did not bend up to the very surface, which meant that inside the crystal, electrons and holes were not affected by any forces, and it made no difference for the carriers to be near the surface or deep in the crystal. Now that we have learnt about the surface states, we realize that things are not so simple. In the case indicated in Fig. 33, the electron is repulsed from the boundary of the crystal and the positively charged hole is attracted to it. The band diagram gives a clear picture of what is taking place. Near the boundary inside the crystal due to a dipole layer a potential barrier is formed, φ_s high. The value φ_s is called a *surface potential*.

For the electrons (the balls on the shaking tray) the barrier presents a small hill (see Fig. 22) from which electrons run down. The higher the hill, the more difficult it is for electrons to climb it, and the fewer electrons climb up to the top of it. If inside the crystal, far from the boundary, the concentration of electrons is n_0, then close to the border, the concentration of conduction electrons will be the following [see Eq. (1)].

$$n_s \approx n_0 \cdot e^{-\varphi_s/kT} . \qquad (27)$$

For the holes which in the band diagram always tend to "flow up" to the surface (see Fig. 18), the region of the barrier near the boundary is a place where holes gather.[a]

The energy of holes near the boundary is smaller than the energy of holes in the volume of the crystal by the value φ_s. Therefore near the surface the hole concentration p_s is

$$p_s \approx p_0 \cdot e^{\varphi_s/kT} , \qquad (28)$$

where p_0 is the concentration of holes deep inside the crystal.

Equations (27), (28) and common sense will help us understand what surface potential on the border of the crystal and the band bends that accompany it may lead to.

Let us investigate, say, the n-type silicon whose concentration of electrons is $n_0 = 10^{15}$ cm^{-3}. Then, according to Table 1 and Eq. (10), the equilibrium concentration of holes in such a crystal is $p_0 = n_i^2/n_0 \approx 10^5$ cm^{-3}. In the volume of crystal there are 10 million times more electrons than holes.

---

[a]Let us again recall the physical sense of it. The field of the surface double charged layer repulses electrons, but attracts holes. Therefore near the surface there are fewer electrons and more holes than inside the crystal volume.

Let the value of the surface potential be $\varphi_s = 0.1$ eV. Then, at room temperature ($kT = 0.026$ eV) the concentration of electrons on the surface is $n_s = 2.10^{13}$ cm^{-3}, and the concentration of holes is $p_s \approx 5.10^6$ cm^{-3}. The number of electrons near the surface is 50 times less than in the volume, the number of holes is as many times more. But the number of holes is still 400 million of times less than that of electrons. Their contribution to conduction is negligible. Therefore the presence of the surface potential in this case implies the appearance near the surface of a depletion layer whose resistivity is much greater than the volume resistivity. As we shall soon see, the appearance on crystal borders of depletion layers with a high resistivity forms the basis of the work of diodes and transistors.

Now let us assume that $\varphi_s = 0.2$ eV. Then $n_s \approx 4.5.10^{11}$ cm^{-3}, $p_s = 2.2.10^8$ cm^{-3}. Qualitatively, the situation is the same. Now on the surface there are more than 2 thousand times less electrons than in the volume, but still they are much more numerous than the holes. Just like it was in the previous case, there is a depletion layer on the boundary. But its width[b] and resistivity is larger than before. But with $\varphi_s = 0.4$ eV, it is a new qualitatively different situation. Now $n_s = 2.10^8$ cm^{-3}, $p_s = 5.10^{11}$ cm^{-3}. The number of holes on the surface is 2.5 thousand times larger than that of electrons! On the surface of the n-type semiconductor, a layer with a hole conductivity has appeared — the layer with the so-called inverse type of conductivity or the *inversion layer*.

We must pay attention to one more circumstance which the presence of surface states leads to. In Fig. 33, it is seen that the presence of the surface potential changes the work function of a semiconductor. In the case shown in Fig. 33, the work function increases by the value φ_s. Indeed, without the surface states the work function of the n-type crystal would approximate the value of the electron affinity χ (see Fig. 30). Now, in order to extract the electron from the crystal volume it is necessary to spend the energy

$$\varphi = \chi + \varphi_s.$$

The band diagram demonstrates that phenomenon quite obviously, as usual. But of course (again as usual), it is necessary to understand clearly the physical cause of the change of work function. In this case, the cause is obvious. In order to fly off from the crystal, the electron, in addition to those

---

[b]We will learn to calculate the depletion layer width a bit further.

forces we have been speaking about when discussing the work function, also has to overcome the repulsion on the surface. The energy necessary to overcome that repulsion is equal to the value φ_s.

There are cases of course when the donor surface states will prevail on the surface of the n-type semiconductor. It is easy to see what that will result in. On giving away electrons, such states will be charged positively and will attract electrons from the volume of crystal (and repulse the holes). As a result, there will appear a layer near the surface of the crystal, enriched with electrons (the accumulation layer). The work function of that crystal will be smaller than in the case when there are no surface states. We suggest that you should draw the band diagram for the n-type semiconductor with a positive surface charge and analyze it the way we did it together for Fig. 33.

If we removed the donor level E_d from Fig. 33, and greatly decrease the number of points electrons and increase the number of circles — the holes, Fig. 33 can be used as a diagram, interpreting the behaviour of the band boundaries in a hole semiconductor whose surface is charged negatively. It is seen that the presence of the surface charge will lead in this case to the formation of an accumulation layer enriched with holes.

And, finally, the fourth case that is possible is the hole semiconductor on whose surface the total charge of surface states is positive. The band diagram for it was given in Fig. 34 and we leave the reader an opportunity to analyze it independently.

Now we are convinced that the presence of surface states can either decrease or increase the work function. And it is certainly the presence of surface states that accounts for the presence of Allen and Gobeli's results (Fig. 32) which seemed paradoxical at first sight. When experimenters split the monocrystal in that section where the donor impurity was prevailing, the n-type traps that appeared on the semiconductor surface greedily captured electrons from the volume of the semiconductor and got the negative charge. As that was taking place, the work function, as we can see it now, was increasing (Fig. 33). The surface states which appeared when the hole semiconductor was being split, captured the holes from the volume in the same greedy way (or, gave their electrons away to the volume), and got a positive charge (Fig. 34).

In both cases, for the n- and p-silicon the value of the surface potential φ_s approximated $E_g/2$. The case when the height of the potential barrier approximates in value $E_g/2$ is met quite frequently. It corresponds to the situation when the majority of surface levels are located near the middle of

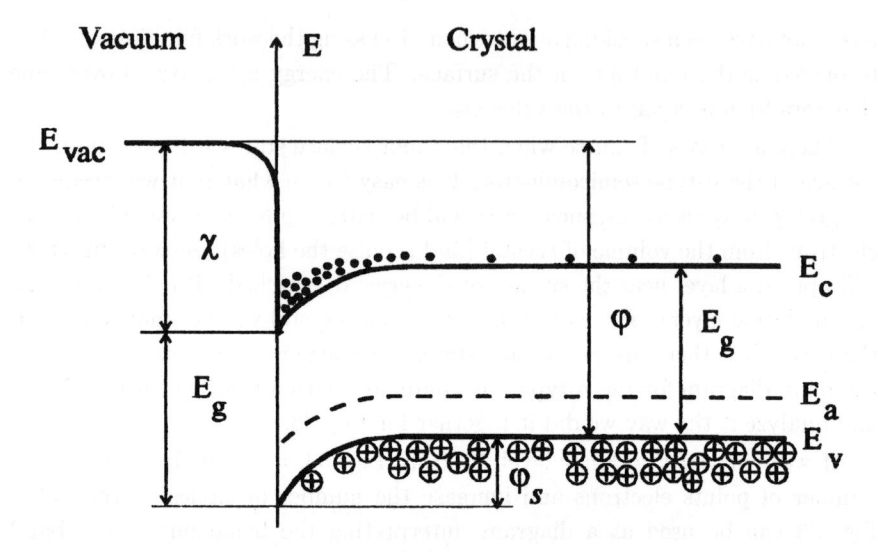

Fig. 34. Band diagram of a p-type semiconductor with positive surface states. Please pay attention that the bending bands in this case decrease the work function as compared to the case when the surface states are absent (Fig. 30).

the semiconductor forbidden band. If $\varphi_s \triangleq E_g/2$, then for the n-type silicon the work function is $\varphi_n \cong \chi + \varphi_s = \chi + E_g/2$ and for the p-type silicon $\varphi_p \cong \chi + E_g - \varphi_s \cong \chi + E_g/2$. Thus $\varphi_n \triangleq \varphi_p$.

3.4 Summary

Near the border line between the crystal and its environment there are forces preventing electrons from leaving the crystal and flying off. The energy necessary for the electron to overcome the effect of these forces (the work function) is several electron volts. In semiconductors, the amount of work function depends on the type of conduction and on the level of semiconductor doping. Besides, it also depends on the spectrum of surface states, i.e. on the surface levels which are always present on the surface of semiconductor. The negative charge of surface states increases the work function, while the positive charge decreases it.

Chapter 4

The Main Parameters of Potential Barriers

Up to now we have been acquainted with only one potential barrier which exists inside the semiconductor — the surface potential barrier. This barrier appears due to the charge of the surface states (Figs. 33 & 34). However, the main parameters are the same for all potential barriers. These parameters are: the potential of the barrier φ (which is often called the height of the barrier), the width of the barrier W and the electric field inside the barrier F.

We have already learnt something about the first parameter. A typical height of the potential barrier φ in semiconductors (and in solids in general) lies in the region from decimal fractions of electron-volt to several electron-volts.

As for the width of the barrier W and the characteristic values of the field F within the barrier, we must note, first of all, that those values are related to each other. With the value of φ being the same, the smaller the width of the barrier (W), the larger the value of F, and vice versa. Indeed, imagine the simplest of the barriers — a barrier whose field is the same everywhere and is equal to F. Then, the width of the barrier being W, the potential drop on that barrier is $U = F \cdot W$. Or, which is the same, the height of that barrier is $\varphi = qU = qF \cdot W$.* Thus, with the value φ given, the value F is $\sim 1/W$.

So, let there be in the semiconductor an energy barrier whose height is φ. What is the width of that barrier?

As we will see, with φ being the same, W can be thousands or even hundreds of thousands of factors smaller or larger, depending on how much the semiconductor is doped. The same enormous difference is characteristic of

---

*To say that between the two points there is a potential barrier whose height is φ is the same as to say that between those points there is a potential difference $U = \varphi/q$. For instance, if between the border of a crystal and its volume spaces there is a potential barrier whose height φ_S is equal to, say, 0.2 eV, that means that between the border and volume of the crystal, there is a potential difference $U_S = 0.2$ V.

the typical values of the field within the barrier F. The simplest way to understand qualitatively why it is so, is to consider the case of the surface barriers. Let us assume that there is an electric field near the crystal surface. As we know, the field can appear either due to the presence of charged surface states or due to the physical mechanisms forming the work function. What interests us is the following: how far will the field penetrate inside the crystal?

In physics it is always easier to consider the problem in its limiting cases. And this is what we are going to do now. First we will see how the electric field penetrates into a metal where there are a lot of free carriers. Then we will see how the field penetrates into a dielectric where there are practically no free carriers. And, in conclusion, we will see how the field penetrates into a semiconductor where, depending on the doping, there may be either very few carriers (a weakly doped semiconductor), or — a lot of carriers (a strong doping).

4.1 How the Electric Field Penetrates Into a Metal, Dielectric and Semiconductor

4.1.1 *Why the electric field practically does not penetrate into a metal*

Let us place a metal plate inside the capacitor the way it is shown in Fig. 35. Though there is an electric field F in the air, between the capacitor plate and

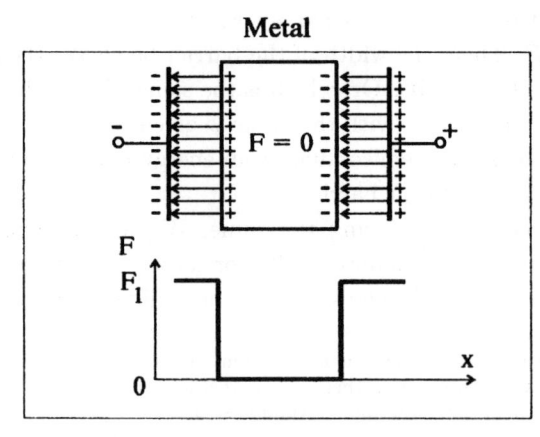

Fig. 35. The electric field F does not penetrate into the metal. Inside the metal, F is equal to zero.

the metal, it does not penetrate into the metal. Inside the metal it is equal to zero. The point is that there are a lot of free electrons in the metal (their concentration being very high $\sim 10^{22}$ cm^{-3}), and exactly the same number of positively charged lattice ions (see Fig. 1). While there is no electric field near the metal, in every volume inside the metal and on the surface of the metal, the number of positive ions is exactly equal to the number of negatively charged electrons. Therefore any part of the metal is electrically neutral.

In the metal, placed into the electric field, the electroneutrality near the surface is broken. Electrons begin to flow to the surface facing the positive contact. And they leave the surface which faces the negative contact, leaving behind them near the surface, a layer of positively charged ions (Fig. 35).

The charge which has thus appeared on the metal surface creates an electric field, directed against the external field. The external field is fully screened, as a result of which the field inside the metal is equal to zero.

The picture shows very distinctly that close to the surface of the metal the electroneutrality is nevertheless broken, and the field does penetrate into the metal, though the depth of this penetration is quite small. How is that depth to be estimated?

Let us assume that under the action of the external field, in order to screen it, all the electrons in the metal plate have left the layer which is Δx thick. And the Δx value is equal just to the lattice constant (which is the distance between the two atoms of the crystal lattice, equal in solids to $\sim 5 \cdot 10^{-10}$ m). What field will then be created near the metal surface? Or, what external field will then be screened?

If all the electrons leave some part of the metal, then inside that part, there will appear a positive space charge whose density will be $\rho = |qN_0|$, where N_0 is the concentration of the positively charged ions, exactly equal to the electron concentration n_0 in the metal. This space charge exists in a thin layer Δx thick. Thus we can say that close to the surface (to the left of Fig. 35) a positively charged plate has appeared whose surface charge density is $\sigma = \rho \Delta x = qN_0 \Delta x$.

From electrostatics, it is known that near the surface with the surface charge density σ, there is an electric field F which is equal to

$$F = \frac{\sigma}{2 \cdot \varepsilon_0} = \frac{q \cdot N_0 \cdot \Delta x}{2 \cdot \varepsilon_0} \qquad (29)$$

(here ε_0 is the dielectric constant). Let us substitute into Eq. (29) the following numerical values: $N_0 = 10^{28}$ m^{-3}, $\Delta x = 5 \cdot 10^{-10}$ m, $\varepsilon_0 = 8.85 \cdot 10^{-12}$ C^2/(N·m^2), $q = 1.6 \cdot 10^{-19}$ C. We obtain: $F = 4.5 \cdot 10^{10}$ V/m!

The value F obtained here is enormously large. And nevertheless in order to create it, it was sufficient just to make the electrons move away from the surface to a distance equal to the thickness of one atom layer.

The same idea can be formulated differently. If a metal is placed into an electric field of enormous intensity, $F \sim 5 \cdot 10^{10}$ V/m, then to have it fully screened, the electrons should move away from the surface to a distance equal to only one lattice constant. If the metal is placed into the field in which the majority of materials would lose their electric stability and there will be a breakdown ($F \sim 10^7 - 10^8$ V/m), then to have such a field screened, it is necessary that only $\sim 0.1\%$ electrons should leave the surface layer. Thus, even the strongest electric fields do not penetrate very deep into the metal, the depth of their penetration is not larger than the lattice constant. And even within that thickness the relative depletion of the electron density on the side of the "minus" of the external field (and, correspondingly, the relative accumulation on the side of the "plus") does not exceed fractions of one percent.

4.1.2 How the electric field penetrates into a dielectric

Figure 36(a) shows a dielectric plate placed between the capacitor plates. Under the action of the external field the dielectric is *polarized*. There appear

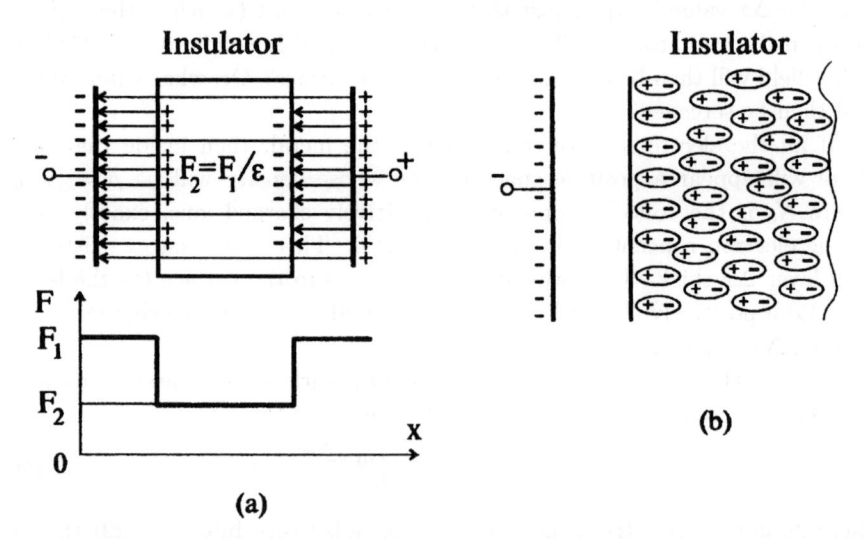

Fig. 36. External electric field is partially screened at the insulator surface. But into the dielectric the field F_2 which has penetrated into a dielectric is the same in every point of the material.

bound charges on the dielectric surface: negative charges on the side of the positively charged capacitor plate, and positive charges on the opposite side. The field of these bound charges is directed against the external field which has generated them. As a result, the external field is partly screened, and the field inside the dielectric becomes smaller than the field outside by a factor of ε (ε is the dielectric constant of a dielectric).

The field F_2 which has penetrated into a dielectric is the same in every point of the material: inside the dielectric there are no free electrons which might rearrange themselves, change their position and screen the action of the electric field which has penetrated into the dielectric.

Figure 36(b) shows what is meant by polarization and why on the dielectric surface the field decreases and in the volume it does not.

There are different types of dielectrics. Some of them can be imagined to be molecules which in the absence of the electric field are in fact "dipoles" — sticks, at one end of which positive charges are concentrated, and at the other, negative charges. In the absence of the electric field, the dipole-sticks are oriented chaotically, their charges being directed in different sides. And in the electric field, the dipoles tend to get arranged in such a way that the positive charge of the dipole should be turned into the "minus", and the negative, into the "plus" of the external field. It is precisely this case that is represented in Fig. 36(c).

It has already been shown to us in the case of metals that in order to screen the external electric field it is necessary that there should be in the material an internal electric field directed against the external field. But take note! Such a field cannot be formed inside the dielectric. Close to every minus of the dipole there is definitely a "plus" of some other dipole. Since there are no free charge carriers (electrons or holes) in a dielectric, and the electric field is unable to break down the charged dipoles, the positive and negative charges cannot part from each other in the volume of the dielectric and form a charged plane.

It is quite different on the dielectric surface. These planes do appear there, as indicated in Fig. 36(a) and (b), and so the field is partly screened (it gets weaker).

Other dielectrics consist of neutral atoms — "balls", comprising a positively charged heavy nucleus and a "light cloud" of negatively charged electrons. If such a dielectric is placed into an electric field, the light electron cloud will stretch in the direction of the "plus" of the external field and a dipole will appear. And then the picture will be exactly the same as that which had been studied by us.

The value ε, which characterizes the weakening of the field at the dielectric border, is called the dielectric constant of a substance. As a rule, the values of ε lie within the range $1 \div 10$. So they are for amber, $\varepsilon \sim 2.8$; for porcelain, $\sim 6 \div 7$; for glass of different makes, $4 \div 10$; for paraffin, $2.2 \div 2.3$. However, there are materials for which the values of ε make hundreds, thousands and even dozens of thousands. They are called *ferroelectrics*.

4.1.3 In what way and how deep the electric field penetrates into a semiconductor

In the semiconductor plate, placed into the external electric field (Fig. 37), on one side, like in a dielectric, there appear bound charges on the surface, due to polarization. Thus, in a semiconductor, at the very border with vacuum (or air) the intensity of the field F_2 will be ε times smaller than in the vacuum (ε is a dielectric constant of the semiconductor). On the other hand, in a semiconductor, just like it is in a metal, there are free charge carriers which are able to rearrange themselves and screen the electric field which has penetrated into the semiconductor.

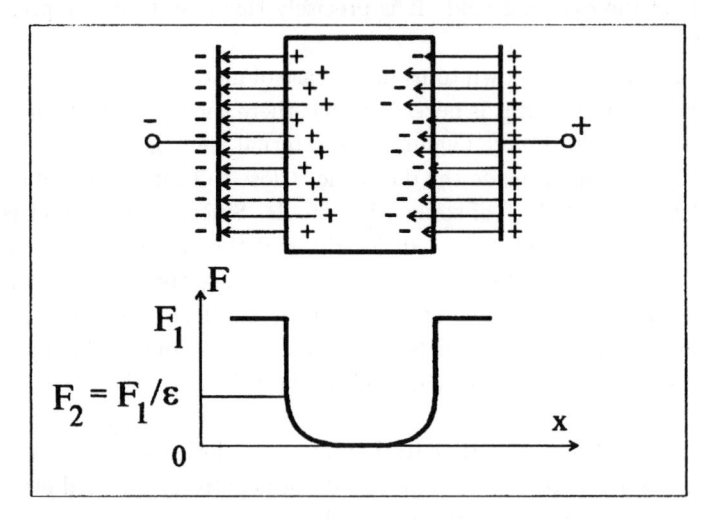

Fig. 37. External field is partially screened at the semiconductor surface (just like in a dielectric). The field F_2 which has penetrated into a semiconductor is also screened. The depth of penetration of the field inside the semiconductor is determined by the level of doping.

In the n-type semiconductor, as we know, there is an equilibrium electron concentration n_0 which is exactly equal to the positively charged donor concentration N_d. While there is no electric field in the semiconductor, in every element of the semiconductor volume, the number of positive ions (donors) is equal to the number of negatively charged electrons. Any part of the semiconductor is electrically neutral.

Furthermore, we could repeat the text from page 89 without changing a single word, beginning with the words "electroneutrality near the surface is broken ..."

Electrons will flow to the semiconductor surface facing the "plus". And they will flow away from the "minus" leaving behind, close to the surface, a layer of positively charged donors.

The charge, created in this way near the semiconductor surface, will set up an electric field directed against the external field which has produced it. The external field is screened, as a result of which the field inside the semiconductor will be equal to zero.

What is then the difference between a semiconductor and a metal? How deep does the field penetrate into a semiconductor? The difference lies in the fact that the concentration of the positively charged ions (donors) and of the negatively charged electrons is much smaller in a semiconductor than it is in a metal. Therefore, to produce a charge large enough to screen the external field, the electrons in the n-type semiconductor must "flow away" from quite a thick layer. The thickness of the layer must be much greater than that of the lattice constant. It is evident that the smaller the level of doping N_d (and, accordingly, the electron concentration $n_0 = N_d$), the greater the thickness of the layer. Thus, the depth of the penetration of the field inside the semiconductor is determined by the level of doping.

Indeed, let us reason the way we did when deriving Eq. (29), but in the "inverse order". Let the field F penetrate into a semiconductor. To screen that field there must be a charged plate near the surface, the density of the surface charge being $\sigma = 2\varepsilon_0 F$. The surface charge density depends on the space charge $\rho = |qN_d|$ which is formed when the electrons have left the surface layer and leave there an uncompensated charge of positively charged donors. Then $\sigma = \rho \cdot \Delta x = qN_d \cdot \Delta x$. Thus, in order to have the field F screened, the electrons must leave the surface layer Δx whose thickness is

$$\Delta x = \frac{2 \cdot \varepsilon_0 \cdot F}{q \cdot N_d} \qquad (30)$$

It is seen that Δx is inversely proportional to the value N_d.

Let the field F, which is to be screened, be, say, $\sim 10^6$ V/m. Then in a semiconductor with $N_d \sim 10^{20}$ cm^{-3}, the width of the surface barrier Δx, according to Eq. (30), will be $\sim 10^{-12}$ m, i.e. 1/500 of the atom layer. (Which means that about $\sim 1/500$ (0.2%) electrons will leave the one-atom surface layer whose thickness is $5 \cdot 10^{-10}$ m). But when the level of doping is $N_d \sim 10^{10}$ cm^{-3}, the width of the surface barrier $W = \Delta x$ is, according to Eq. 30, about 10^{-2} m, i.e. 1 cm or 20 000 000 atom layers.

Up to now we have been discussing only an electron semiconductor. But the same refers to a hole semiconductor. As we know, in the p-type semiconductor there is an equilibrium concentration of positively charged mobile carriers - holes, exactly equal to the concentration of the fixed negatively charged acceptors N_a. If the p-type semiconductor is placed into an external electric field, the holes will flow up to the semiconductor surface, facing the negative electrode of the external field. And the uncompensated space charge of the negatively charged acceptors will stay near the semiconductor surface facing the positive electrode. The smaller the acceptor concentration, i.e. the lower the level of doping the semiconductor with an acceptor impurity, the thicker the layer must be, so as to produce the necessary charge to screen the external field.

4.2 Field Dependence on the Coordinate

So, we have established that as a result of screening, the electric field falls deep into the semiconductor and exists only at a distance of $W = \Delta x$ from the surface, with the value W being inversely proportional to the level of doping N. It is pertinent to ask: How, according to what law, does the field decrease deep inside the semiconductor?

Let there be a surface potential barrier in a semiconductor (Fig. 33). Near the surface the field is maximal and is equal to F_m. At the distance W from the surface, the field is equal to zero. The simplest assumption apparently is that the field decays deep inside the semiconductor the way it is shown in Fig. 38, i.e. linearly. The higher the doping level, the more abruptly the field decays and the smaller is W, with the value F_m being the same.

The most striking thing is that in the majority of cases which are practically interesting to us this simplest supposition proves to be correct. At the end of this section, we will prove it quite rigorously. And meanwhile we will trust it to be so, i.e. admit that the field F decreases with the x coordinate just according to that law and that the width of the barrier W

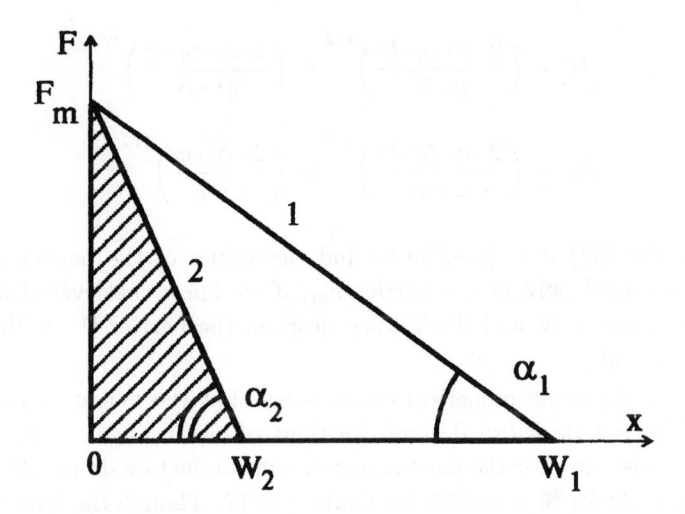

Fig. 38. In the majority of cases which are of a practical interest, the field decays deep inside the semiconductor linearly. The higher the doping level, the greater is the field slope: $\tan \alpha = \frac{F_m}{W} = \frac{q \cdot N}{\varepsilon \cdot \varepsilon_0}$.

and the maximal field F_m are connected with each other by the relation of the validity which will be proven later:

$$W = \frac{\varepsilon \cdot \varepsilon_0 \cdot F_m}{q \cdot N} \qquad (31)$$

where ε is the dielectric constant of the semiconductor; N is the impurity concentration.

It is easy to verify (see Fig. 38) that Eq. (31) corresponds to the condition

$$\tan \alpha = \frac{F_m}{W} = \frac{q \cdot N}{\varepsilon \cdot \varepsilon_0}$$

And now, believing Eq. (31) to be valid, we will establish a few important relationships. If there is a barrier in which the distribution of the field $F(x)$ is known, the voltage drop on the barrier U is known to be equal to the area under the curve $F(x)$.[a] For the distribution $F(x)$, shown in Fig. 38, the value U can be calculated quite easily: $U = 1/2 \cdot E_m \cdot W$. Or, otherwise, using Eq. (31):

---
[a] Or, which is the same, $U = \int_0^W F(x) \cdot dx$

$$W = \left(\frac{2 \cdot \varepsilon \cdot \varepsilon_0 \cdot U}{q \cdot N}\right)^{1/2} \equiv \left(\frac{2 \cdot \varepsilon \cdot \varepsilon_0 \cdot \varphi}{q^2 \cdot N}\right)^{1/2}$$

$$F_m = \left(\frac{2 \cdot q \cdot N \cdot U}{\varepsilon \cdot \varepsilon_0}\right)^{1/2} \equiv \left(\frac{2 \cdot N \cdot \varphi}{\varepsilon \cdot \varepsilon_0}\right)^{1/2} \tag{32}$$

Using Eq. (32) it is possible to find the values of the barrier width W and the maximal field in the barrier F_m, if we know the level of doping of the semiconductor N and the voltage drop on the barrier U (or the barrier potential $\varphi = qU$).

Let us make some numerical estimations. Let the height of the surface potential barrier be a few decimal fractions of electron-volt. The values of dielectric constant ε for the most common semiconductors do not differ much. For Ge, $\varepsilon = 16$; for Si, $\varepsilon = 12.5$; for GaAs, $\varepsilon = 12$. Though the level of doping can vary in semiconductor devices within very wide ranges: from $\sim 10^{10}$ to $\sim 10^{20}$ cm^{-3}. Let N be equal to 10^{10} cm^{-3}. We will assume, to make it more clear, that on the surface of GaAs there is a barrier whose height is $\varphi_S = 0.2$ eV ($U_s = 0.2$ V). Then, substituting the values $\varepsilon = 12$ and $N = 10^{16}$ m^{-3} into Eq. (32), we obtain $F_m = 2.4 \cdot 10^3$ V/m=24 V/cm, $W = 1.7 \cdot 10^{-4}$ m $= 1.7 \cdot 10^{-2}$ cm. But if there is the same surface potential difference $U_S = 0.2$ V on the surface of a highly doped GaAs plate with $N = 10^{20}$ cm^{-3}, then $F_m = 2.4 \cdot 10^6$ V/cm, $W = 1.7 \cdot 10^{-7}$ cm. In the first case the field of the surface barrier penetrates into the semiconductor as deep as 300 000 atom layers. In the second case, of the order of 3-4 layers.

So far we have been discussing the n-type semiconductors, with electrons as mobile carriers. It goes without saying that if we deal not with an electron — but with a hole semiconductor, all the above calculations and all the derived equations are still valid. But it will be the positively charged holes that will move under the field, leaving the negative charge of the acceptors uncompensated.

4.3 Poisson's Equation

In this section, we will derive (together with you) one of the basic equations of the semiconductor physics. This equation connects the change of the field with the coordinate $\frac{dF}{dx}$ with the doping of a semiconductor. It has the following form:

$$\frac{dF}{dx} = \frac{\rho}{\varepsilon \cdot \varepsilon_0} \tag{33}$$

Here ρ is the space charge density.

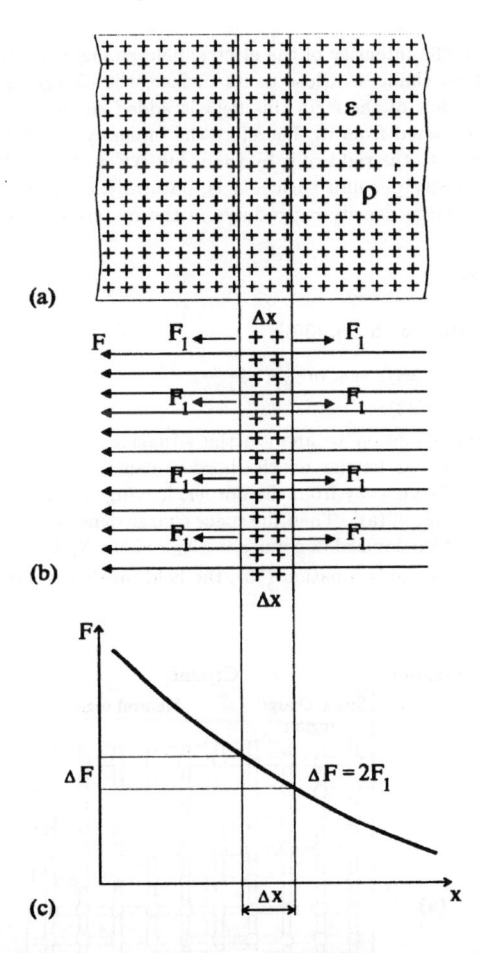

Fig. 39. To the left of the charged strip the field F_1 is added to the external field F, and to the right of it, it is subtracted from the field F. Within the strip Δx thick the intensity of the electric field changes by the value $\Delta F = 2 \cdot F_1$.

In physics of semiconductors Eq. (33) is often called Poisson's equation.

By deriving this equation we will prove the validity of Eq. (31) and establish the cases in which the field F decreases linearly with the x coordinate.

Let us study very attentively the part (region) of the semiconductor containing the space charge [Fig. 39(a)]. Let this charge be positive. (To make it more impressive we can picture this region as the left part of a semiconductor plate, shown in Fig. 37. Electrons have flown away from that part of the plate, leaving behind them the donor's positive charge with the space density ρ). The field in this region changes. Our task is to define the law of the field change. We will specify within the region under consideration a narrow strip which is Δx

thick. Within this strip the intensity of the electric field changes by the value ΔF. We can understand why the field changes. Close to its surface, the charged strip produces a field F_1 [Fig. 39(b)]. To the left of the strip this field is added to the external field F, and to the right of it, it is subtracted from it. Therefore the intensity of the field to the left of the strip will be greater than to the right of it by the value $\Delta F = 2 \cdot F_1$. With the space charge density ρ the charge density per unit surface σ is, naturally, equal to $\sigma = \rho \cdot \Delta x$. And the strip with the surface charge density σ produces in the material with a dielectric constant ξ_0 field F_1, equal to

$$F_1 = \frac{\sigma}{2 \cdot \varepsilon \cdot \varepsilon_0} = \frac{\rho \cdot \Delta x}{2 \cdot \varepsilon \cdot \varepsilon_0}$$

[Compare this equation with Eq. (30)].

Thus, $\Delta F = 2F_1 = \rho \cdot \Delta x / \varepsilon \cdot \varepsilon_0$, or $\frac{dF}{dx} = \frac{\rho}{\varepsilon \cdot \varepsilon_0}$.

We have obtained Eq. (33).

Let us now use this equation to analyze the situation that is already well-known to us. Let there be a potential barrier on the border, caused by the fact that the mobile carriers (electrons) have flown away from the border, leaving the space charge of the ionized donors uncompensated [Fig. 40(a)]. Then the space charge density ρ is naturally equal to the uncompensated charge of the ionized donors: $\rho = qN_d$, where N_d is the donor concentration.

In accordance with Poisson's equation (33), the field inside the surface layer will change according to the law

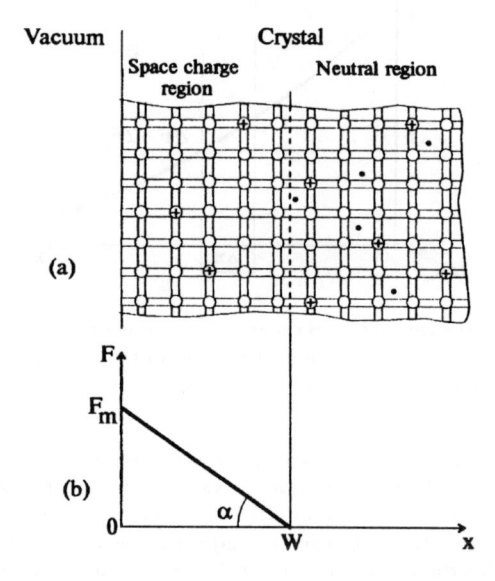

Fig. 40. Formation of the surface potential barrier (the n-type semiconductor): (a) mobile carriers (electrons) have flown away from the border due to the negative charge of surface states, leaving the space charge of the ionized donors uncompensated; (b) the field F is linearly falling with the coordinate x near the surface.

$$\frac{dF}{dx} = \frac{\rho}{\varepsilon \cdot \varepsilon_0} = \frac{q}{\varepsilon \cdot \varepsilon_0} \cdot N_d \qquad (34)$$

Let us assume that the doping of a semiconductor is homogeneous, i.e. the value N_d is the same everywhere. Then, in accordance with Eq. (34) the field F is linearly falling with the coordinate x (Fig. 40(b)).

The slope α is defined by Eq. (34):

$$\tan \alpha = \frac{dF}{dx} = \frac{q}{\varepsilon \cdot \varepsilon_0} \cdot N_d$$

In Fig. 40(b) we can see that $tg\,\alpha = F_m/W$. Thus

$$W = \frac{F_m}{tg\,\alpha} = \frac{\varepsilon \cdot \varepsilon_0 \cdot F_m}{q \cdot N_d}$$

We have obtained Eq. (31). And thus we have proved the validity of Eq. (32).

4.4 A Few Words about Accumulation Layers

A natural question may arise: we have been discussing all the time the width and other parameters of the depletion layers. Meanwhile, as we know, there also exist accumulation layers, rich in carriers (electrons or holes). Why then do we show so much interest in the former, neglecting the latter?

When we derive Poisson's equation (33), it does not matter to us what the nature of space charge ρ is. It makes no difference whether it was brought about due to the shortage or to the excess of carriers in the semiconductor. So Eq. (33) can be used very well to calculate the parameters of either depletion or accumulation layers.

Our greater interest in the depletion layers has two grounds. First, it is these layers that we will mostly deal with when analyzing the work of semiconductor devices. Secondly, it is much simpler to calculate the parameters of the depletion layers. The reason is quite clear. Being poor in carriers, they have quite an evident natural limit. If all the carriers have left, that means that there are no more of them present. And if some part of a semiconductor has no carriers, then the space charge density ρ in it is equal either to qN_d (the n-type semiconductor), or to $-qN_a$ (p-type).

When the layer is rich in carriers, things are more complicated. At first sight it may even seem ambiguous if having an excess of carriers implies having a natural limit. However, the accumulation of carriers is limited by the diffusion process known to us. The greater the concentration of the free carriers, the greater the diffusion flux of carriers, opposing their further accumulation [see Eq. (17)].

The space charge distribution has a more complicated dependence on the coordinate when there is an excess of carriers rather than a lack.

However the qualitative conclusion remains valid for the accumulation layers as well. The lower the level of doping, the larger the width of the layers and the smaller the field amplitude, the value of the potential barrier being the same.

4.5 Summary

The width of the potential barriers in a semiconductor is determined by the level of doping. For the weakly doped semiconductors it may make hundreds of thousands of atom layers (tens of micrometers); for the high doped semiconductors — several atom layers (thousands fractions of micrometre). When the barrier height is given: $\varphi = qU$, its width W is inversely proportional to $N^{1/2}$ (N is the concentration of the doping impurity), and the maximal field within the barrier F_m is directly proportional to $N^{1/2}$.

Chapter 5

p-n Junction

> ... No, there never has been and never will be any wisdom more trustworthy than that which I'll tell you now, oh mullah! But get ready not to be struck by it too hard, for it is easy to go out of one's mind on hearing it — so striking, dazzling and unbounded it is.
>
> L. V. Solovyev
> *"Story of Hodzha Nasredin"*.

The junction we are going to study in this chapter has been under investigation for more than half a century. All its properties have been studied in detail. Dozens of various semiconductor devices have been made on its basis. It seems there is no such question, connected with this junction, the answer to which has not been obtained. And yet every year either one or several new devices come into being which are based on the properties of the same *p-n* junction.

If a semiconductor crystal is doped in such a way that one part of it has a *p*-type (hole) conductivity, and the other part has an *n*-type (electron) conductivity, then on the boundary between these two parts there appears a layer with quite specific properties, which is called a *p-n* junction.

5.1 Ways of Obtaining *p-n* Junctions

The first idea that strikes an inexperienced man, pondering on how to produce a *p-n* junction, consists, as a rule, in the following. One has to take two pieces of semiconductor: one of a *p*-type, the other of an *n*-type, polish them thoroughly and press firmly together.

Half a century ago, when the semiconductor technology was in its infancy, specialists even tried to put the above into practice by means of numerous experiments. Alas, that simple way proved to be useless, as a rule. Should we do all those things, we would obtain an object whose properties would have nothing in common with those of the p-n junction which we are going to study. The real p-n junction, whose properties we are interested in, is shown in Fig. 41. Figure 41(a) shows it schematically, and Fig. 41(b) shows it in detail. Pay attention to a most important circumstance: the crystal lattice of a semiconductor does not know anything about the p-n junction, it merely does not notice it. Indeed, on the boundary dividing the p and n regions, to the left or to the right from it, we see the same silicon crystal lattice everywhere, not affected at all by different impurities introduced into different parts of the crystal. Only such a junction possesses all those wonderful properties we have been speaking about. And it is this junction that we are going to study.

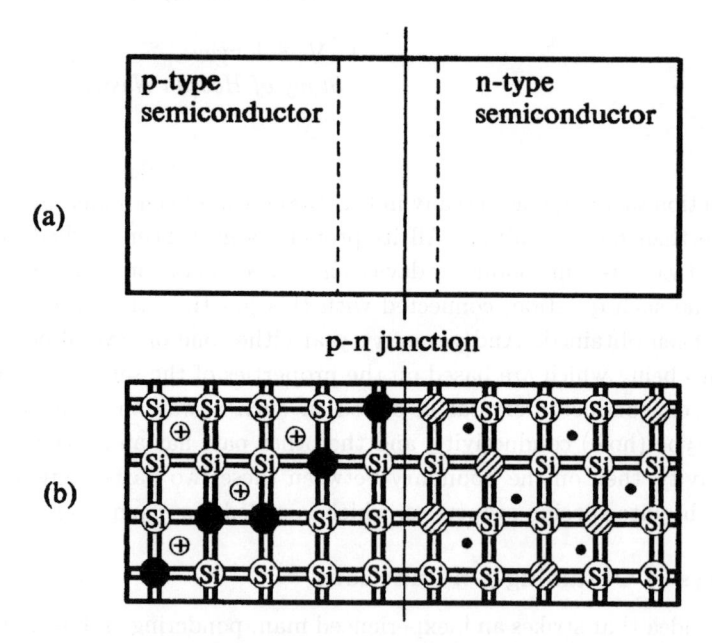

Fig. 41. (a) A p-n junction appears on the boundary of n- and p-types of the same semiconductor; (b) As an example a Si crystal is shown. In the left part of the crystal, the acceptor impurity is introduced (dark circles); in the right part of the crystal, the donor impurity is introduced (shaded circles).

Meanwhile, having read Chapter 3, we understand very well that on the boundary of either of the two pieces of the semiconductor of p and n type, no matter how thoroughly their surface might have been treated, the crystal lattice might be destroyed. A spectrum of surface states is formed. Surface potential barriers will appear. In a word, the same phenomena which we have studied in Chapter 3.

No matter how hard you press pieces of p and n type of semiconductors together, no matter how thoroughly their surface has been polished, the injury caused to the surface of the lattice cannot be cured. The crystal lattice on the boundary will always be broken: there will always be oxides and absorbed foreign atoms.

For more than half a century specialists in many countries have been trying their best to solve the problem of creating p-n junctions of a good quality without either those or any other defects. No wonder there are a lot of methods of obtaining such junctions, all those methods being technically quite perfect.

The methods we are going to consider now can be characterized by one common expedient. One takes a semiconductor crystal of p or n type. Then in one way or the other some impurities are introduced to a certain depth of the crystal, forming in this crystal an impurity of the opposite character. If initially it was an n-type crystal, the acceptor impurity is introduced; if it was a p-type crystal, the donor impurity is introduced. The concentration of the impurity that is being introduced should be greater than that of the initial crystal. Therefore, in that part of the crystal where the impurity was introduced, the type of the conductivity is changed. The so-called *overcompensation* of the initial impurity is taking place.

A p-n junction appears on the boundary between the region where the type of the conductivity has changed and that where it has remained initial.

5.1.1 *Alloying*

Alloying is one of the oldest ways of forming a p-n junction. The first p-n junctions were obtained in this way in 1950. The method is based on forming an alloy of a semiconductor material with a metal whose atoms are donors or acceptors in the semiconductor.

Figure 42(a) shows, as an example, the formation of a alloyed p-n junction in n-Ge when indium (In) is being alloyed into it. A tablet of indium in the form of a thin disc is placed on a clean surface of germanium crystal. Then the crystal together with the tablet on it is heated very slowly. As soon as the temperature exceeds that of melting of the metallic tablet — for indium it is 156°C — the tablet begins to melt, becoming a drop of liquid. When the temperature rises, the shape of the drop changes. Forces of the surface tension of liquid are striving to shape the drop into a ball, and forces of wetting tend to spread the drop on the surface of the crystal. Indium which has been made into a ball at $\sim 300°C$, spreads over the surface of germanium at higher temperatures ($\sim 500°C$).

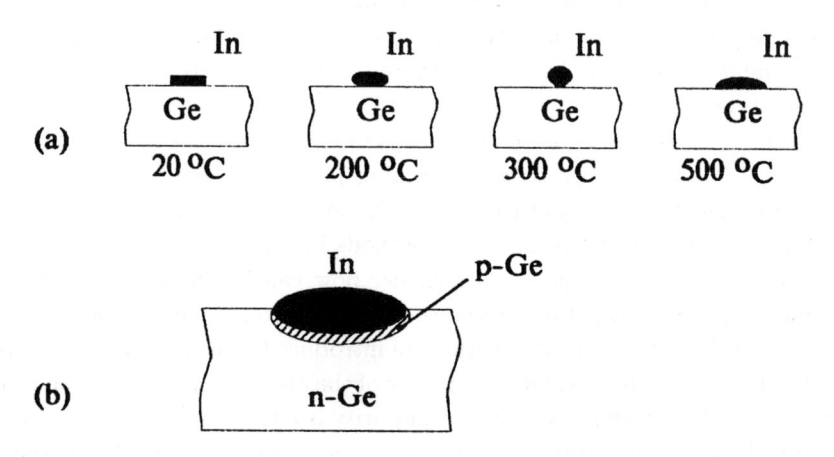

Fig. 42. Formation of an alloyed p-n junction in n-Ge. (a) the change of the In tablet shape with the increase of temperature; (b) the recrystallised layer which has been formed on the boundary between Ge and In proves to be enriched with atoms of indium.

Crystalline germanium dissolves very well in liquid indium at temperature $\geq 500°C$. Dissolution goes on until a saturated solution of germanium in indium is formed. After this, the germanium plate is slowly cooled. With the fall in temperature the amount of germanium that can dissolve in indium decreases; the excess of germanium, precipitated from the solution, crystallizes again (*recrystallizes*) on the part of the tablet that is undissolved. The recrystallized layer which has been formed [Fig. 42(b)] proves to be enriched with atoms of indium.

In germanium, atoms of indium are acceptors. The concentration of indium in the recrystallized layer makes $\sim 10^{19}$ cm^{-3} and greatly exceeds the concentration of the donor impurity in the initial crystal. That is why a p-n junction is formed on the boundary of the recrystallized zone and of the undissolved part of the crystal.

Due to relative simplicity and high reliability, this method is still being used when manufacturing semiconductor devices.

5.1.2 *Diffusion*

Discussing phenomena of diffusion in Chapter 2, we have already mentioned that due to the process of diffusion, impurity atoms can penetrate from the surface into the bulk of the semiconductor. Now we will consider this process in detail.

Let us assume that a semiconductor crystal is placed into the atmosphere of gas, containing high concentrations of atoms of a substance which acts in this semiconductor as a donor (or acceptor) impurity. As we know, on the surface of any crystal the crystal lattice is broken in many places. So the impurity atoms of the gaseous diffusant are easily captured by the surface of the semiconductor. The surface appears to be enriched with the impurity. And what comes then? How can the impurity penetrate from the surface into the bulk of the crystal?

We are not dealing here with electrons or holes — charge carriers able to move freely inside the crystal. We are discussing the behaviour of atoms. First, atoms of the impurity as such are fixed on the surface of the crystal. And secondly, which is perhaps more important, atoms of the semiconductor, which are to be substituted by the impurity atoms, are also fixed (and fixed very fast) in the sites of the crystal lattice. How, under these conditions, can diffusion take place?

Indeed, there is practically no diffusion in the absolutely perfect crystal at a temperature of absolute zero.

But absolutely perfect crystals do not exist. In any real crystal, even at low temperatures, there are always defects in the crystal lattice. And if the crystal is heated, the thermal oscillations of the lattice will result in the fact that some of the semiconductor atoms will jump off from their proper places in the sites of the lattice. Another atom, including that of the impurity, can jump into the vacated place (*the vacancy*). Atoms, both intrinsic and impurity, driven away from their places in the lattice into the interstices, can travel about the crystal much more easily. Thus, the diffusion of impurity goes on mainly in the vacancies, interstices and other defects of the semiconductor crystal lattice.

The diffusion flow of the impurity atoms is naturally directed from the surface, where there are many impurity atoms, into the depth of the semiconductor, where there are no impurity atoms at all. The value of the diffusion flux, i.e. the quantity of diffusion atoms, crossing a unit of area per unit time is determined by Eq. (17).

But the value of the diffusion coefficient D, determining the velocity of the atom diffusion is, as a rule, millions of billions of times less than the coefficient of diffusion of electrons or holes.

With the temperature rises, the impurity diffusion coefficient increases very abruptly, exponentially (Fig. 43).

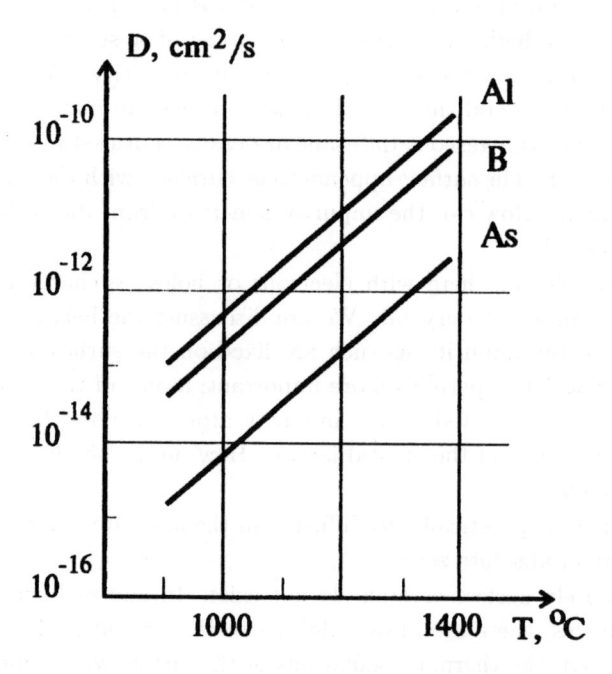

Fig. 43. Temperature dependences of the impurity diffusion coefficient in Si.

The distance, which in the process of diffusing the impurity atoms have covered inside the semiconductor, can be roughly estimated by Eq. (24), provided the value of the diffusion coefficient at the temperature of the process (Fig. 43) and the diffusion time are known.

If there is, say, a diffusion of the acceptor impurity into the n-type semiconductor, then in the region where after the diffusion the acceptor impurity will be greater than the donor impurity, the semiconductor will be of a hole type (*p*-type semiconductor). In the regions which have not been reached by the impurity atoms in the process of diffusion, the semiconductor will remain of an electron type (*n*-type). A *p-n* junction will appear on the boundary between these two regions (Fig. 44).

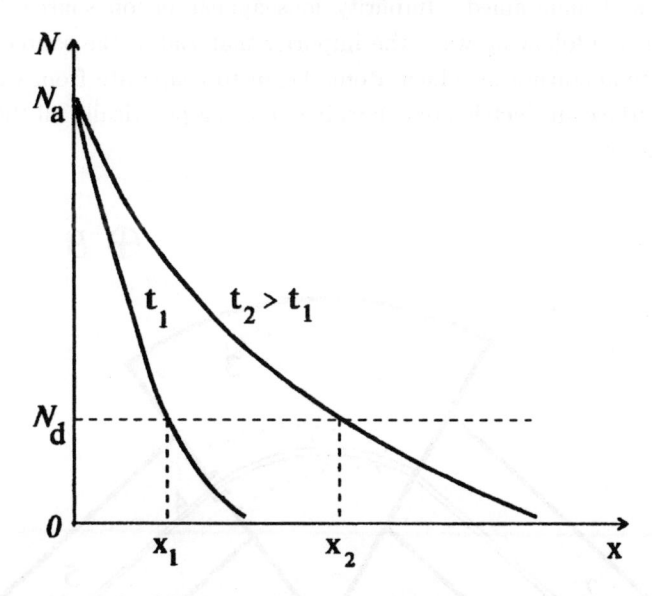

Fig. 44. Formation of the diffusion *p-n* junction. Acceptor impurity is introduced into the *n*-type semiconductor with the donor concentration N_d. A high concentration of acceptor impurity N_d is created on the surface of the semiconductor plane. Due to the diffusion process, the acceptor impurity penetrates into the depth of the wafer. The longer the diffusion time t and the higher temperature T, the greater is the depth x at which the *p-n* junction occurs.

The depth at which the *p-n* junction occurs, i.e. the distance from the *p-n* junction to the surface of the crystal can vary very much from a fraction of a micron to hundreds of microns.

5.1.3 *Ion implantation*

Ion implantation is one of the most modern ways of producing *p-n* junctions, which makes it possible to obtain *p-n* junctions of very small sizes, with an exactly controlled and very small (to a hundredth of a fraction of a micrometre) depth. The method of ion implantation, or, as it is sometimes called, of ion doping, consists in introducing into a semiconductor the atoms which do not rush chaotically in all directions, as they do when diffusion takes place, but move in one and the same direction with the same, and rather high, velocity.

The installation for producing an ion implantation is schematically shown in Fig. 45. In chamber 1 where the process takes place, a deep vacuum is created and maintained. Impurity ions appear in ion source 2 which is arranged in the following way: the impurity material in the source is heated up to the temperature at which atoms begin to evaporate from the surface. Near the surface an electric arc is burning, ionising practically all the particles that get into it.

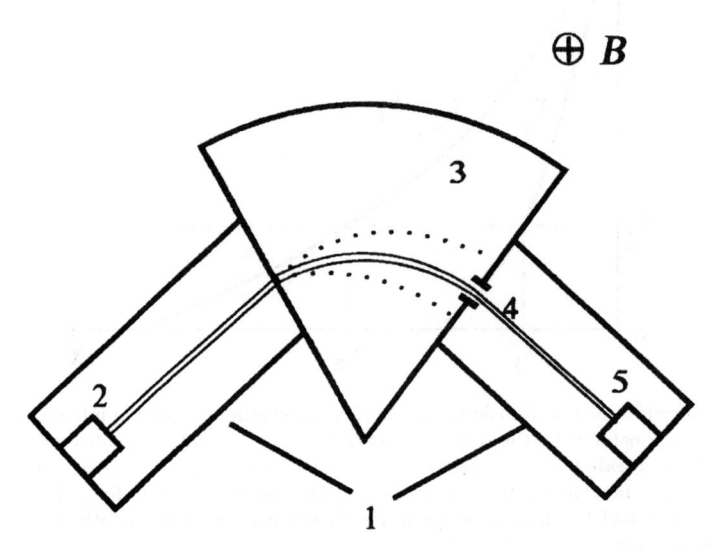

Fig. 45. The installation for producing an ion implantation. 1 — the chamber; 2 — the ion source, 3 — the magnetic analyser with an outlet slot 4; 5 — a semiconductor wafer ("target").

The positive impurity ions which have been formed are driven away through the outlet of the source by a strong electric field. A special system of electrodes (called "electrostatic lens") gives the beam the necessary needle-like shape.

The focused ions, accelerated by the field to a great velocity, get into magnetic analyser 3, a most important nude which provides one of the main advantages of the method of ion doping: a practically complete clearing of the ion beam from any foreign impurities.

Chapter 5. p-n Junction 107

In the analyser the ion beam gets into a transverse magnetic field (lines of the magnetic field B in Fig. 45 are perpendicular to the plane of the figure). As it is known, the flying charged particles deviate in the magnetic field under the action of the Lorentz force. The radius of the trajectory of ions which got into the magnetic field depends on their velocity and on their mass. So, should there be foreign particles in the beam, they will deviate in the analyser by the angle not equal to that at which the impurity ions will deviate.

There is a slot in the outlet wall of the analyser through which only those ions which have deviated to a strictly given angle pass. It is quite easy to "choose" the type of ions in a beam: it is enough to change the magnetic field of the analyser.

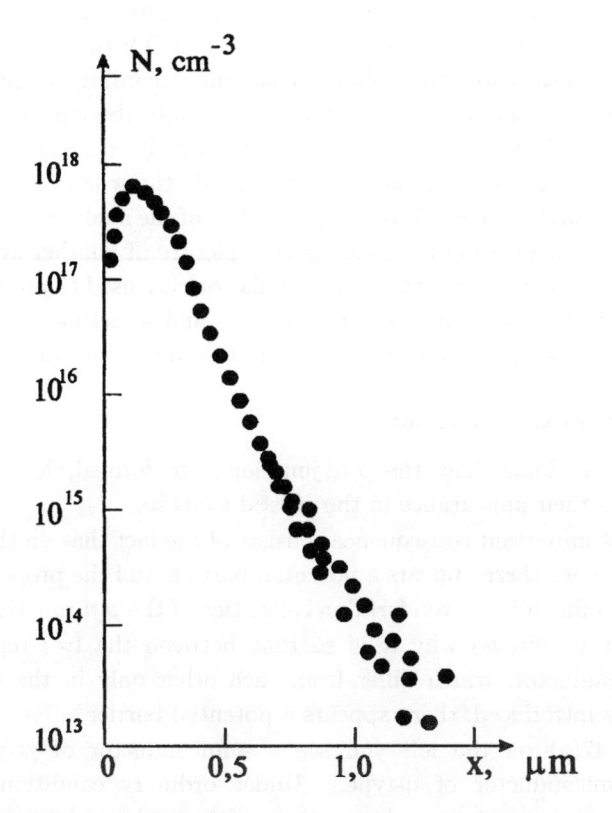

Fig. 46. An example of the distribution of the impurity introduced into the crystal by ion implantation.

The "filtered off" impurity ions which have flown from the analyser fly up to a semiconductor plate. Colliding with the semiconductor lattice, the ions gradually lose their energy, giving off to the atoms they come across on their way and finally they stop. The distance the ion will have covered until it stops, or the depth of its implantation, depends on the mass and energy of the ion. The energy of ions can be controlled quite accurately by changing the electric field, accelerating the ions. Thus, the implantation of the impurity ions into the lattice goes to a strictly specified depth.

Ion implantation is one of the most exact methods of producing p-n junctions (Fig. 46). Any wanted impurity can be implanted by this method into a semiconductor in a strictly controlled quantity and to a given depth, all the foreign impurities being filtered off.

In some installations of ion implantation, electrostatic deviating plates are placed on the way of the ion beam, just like those in TV tubes or oscilloscopes. Applying to these plates the voltage, changing according to the given law, generated by the computer, it is possible to deviate the ion "needle" to the wanted point of the plate. Thus, the ion beam directly "depicts" the necessary configuration of the p-n junctions. Having made the whole "picture" of one impurity, and having changed the magnetic field of the analyser and the energy of the beam, it is possible to make another picture of another system of p-n junctions at another depth, etc. Such installations are used to produce integral circuits in which thousands, tens of thousands and sometimes even hundreds of thousands of p-n junctions are located on the same semiconductor plate.

5.2 Barrier on the Boundary

Now, when we know how the p-n junctions are formed, let us see what consequences their appearance in the crystal leads to.

The most important consequence consists of the fact that on the boundary of the p-n regions there appears a potential barrier, and the properties of that barrier determine all the wonderful peculiarities of the p-n junction.

First, let us discuss why it is so that between the two regions of the same semiconductor, which differ from each other only in the character of the impurity introduced, there appears a potential barrier.

In Fig. 47(a) on the left you see a semiconductor of p-type; to the right, a semiconductor of n-type. Under ordinary conditions, the hole semiconductor contains, as we know, the positively charged particles — holes — mobile particles, able to conduct the current, and the negatively charged acceptors, fixed steadily in the crystal lattice. The electronic semiconductor, on the contrary, contains negatively charged conduction electrons and positively charged ionized donors, steadily fixed in the lattice.

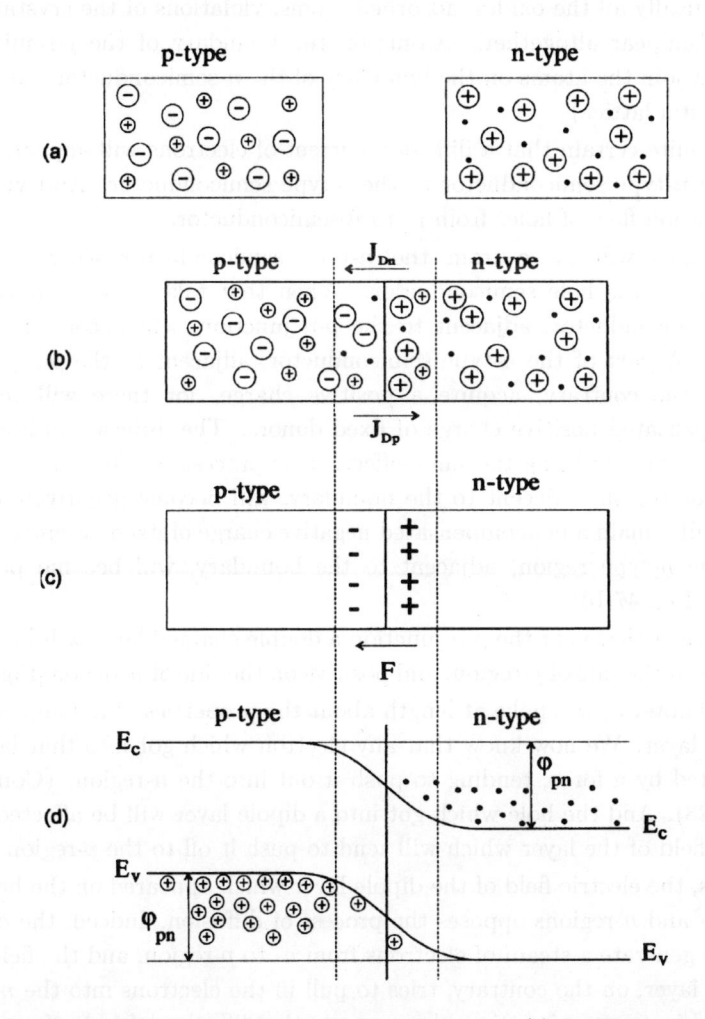

Fig. 47. Formation of the potential barrier of the *p-n* junction. (a) the *p*-type semiconductor contains positive mobile particles (holes), the *n*-type semiconductor contains negative mobile particles; (b) electrons diffuse into the *p*-type semiconductor and hole diffuse into the *n*-type semiconductor; (c) a dipole layer appears at the boundary between the *n*- and *p*-semiconductors; (d) the presence of the dipole layer is equivalent to the existence of the energy (potential) barrier.

Figure 47(b) shows what would happen if we suddenly became magicians and managed to join these two pieces of the semiconductor together. (Not just by pressing them to each other, but by joining them in such a way that magically all the oxides, adsorbed atoms, violations of the crystal lattice might disappear altogether. Atoms on the boundary of the p-semiconductor might join the atoms on the boundary of the n-semiconductor, forming an ideal crystal lattice).

It is quite certain that a diffusion current of electrons will appear, flowing from the n-type semiconductor to the p-type semiconductor. And vice versa — a diffusion flow of holes from p- to n-semiconductor.

Electrons will move from the n-type semiconductor where they are numerous to the hole semiconductor. When that takes place a part of the p-type semiconductor, adjacent to the p-n junction, will become negatively charged. A part of the n-type semiconductor, adjacent to the p-n junction, will, on the contrary, acquire a positive charge, for there will remain a noncompensated positive charge of fixed donors. The diffusion of holes from p- to n-region will bring the same effect. This process results in the fact that the p-type region, adjacent to the boundary, will become negatively charged (there will remain a noncompensated negative charge of fixed acceptors there), while the n-type region, adjacent to the boundary, will become positively charged [Fig. 47(b)].

It is clear that near the p-n junction a double charged layer will be formed, negative on the side of p-region, and positive on the side of n-region [Fig. 47(c)].

In Chapter 3, we spoke at length about the properties of a double-charged (dipole) layer. We now know that any electron which got into that layer will be affected by a force, tending to push it out into the n-region. (Compare it to Fig. 28). And the hole which got into a dipole layer will be affected by the electric field of the layer which will tend to push it off to the p-region.

Thus, the electric field of the dipole layer which appeared on the boundary of the p- and n-regions opposes the process of diffusion. Indeed, the diffusion tends to generate a steam of electrons from n- to p-region, and the field of the charged layer, on the contrary, tries to pull in the electrons into the n-region. The field in the p-n junction also opposes the diffusion of holes from p- into n-junction in exactly the same way.

In the end, as a result of the two processes, acting in the opposite directions: the diffusion and the motion of charges in the electric field, a steady state is established which does not depend on time.

In this steady state the diffusion stream of electrons from the right to the left is exactly compensated by the stream of electrons in the electric field F (the drift current) from the left to the right. Of course, the same refers to holes: the diffusion stream of holes from p- into n-region is absolutely compensated by the drift current flowing in the opposite direction.

So, on the boundary between the p- and n-regions there is a dipole layer which opposes the penetration of electrons from n- into p-semiconductors, and the penetration of holes from p- into n-semiconductors. But as we have learnt from Chapter 3, the presence of the dipole layer is equivalent to the existence of the energy (potential) barrier. [Fig. 47(d)].

The band diagram in Fig. 47(d) shows quite clearly that in order to get from n-region into p-region, the electron must overcome the energy barrier whose height is φ_{pn}. The hole must also overcome a similar barrier in order to get from p- into n-region of the p-n junction.

5.2.1 *The height of the barrier*

To define the height of the barrier, i.e. the value of φ_{pn}, let us discuss once again why on the boundary between the p- and n-regions there appears an energy barrier. But this time we will discuss this problem from the very beginning in the language of energy diagrams.

Figure 48(a) shows the diagrams that are already quite familiar to us (compare this with Fig. 30): on the left for p-semiconductors, on the right for n-type semiconductors. If on the boundaries of p- and n-crystals there are no surface states or no surface charge connected with them, then the work function of the n-type material φ_n is less than that of the p-material φ_p by the value E_g. That circumstance that φ_n is essentially less than φ_p means that it is much easier for the electrons to get outside (into the vacuum) from n-material than from the material of p-type.

Let us now bring together the crystals of p- and n-types, so that they might exchange the electrons that have flown outside. [Fig. 48(b)]. Two flows of electrons will arise, directed towards each other: a flow of electrons J_{nn}, directed from an n-type crystal to a p-type crystal, and a flow of electrons J_{np}, from a p-type crystal to an n-type crystal. It is also clear that $J_{nn} > J_{np}$.

Since more electrons pass into the p-material than into the n-material, it is clear that the p-region will be charged negatively as related to the n-region. This circumstance will hinder any further transition of electrons from the n- into the p-region, and eventually a steady state will be established, at

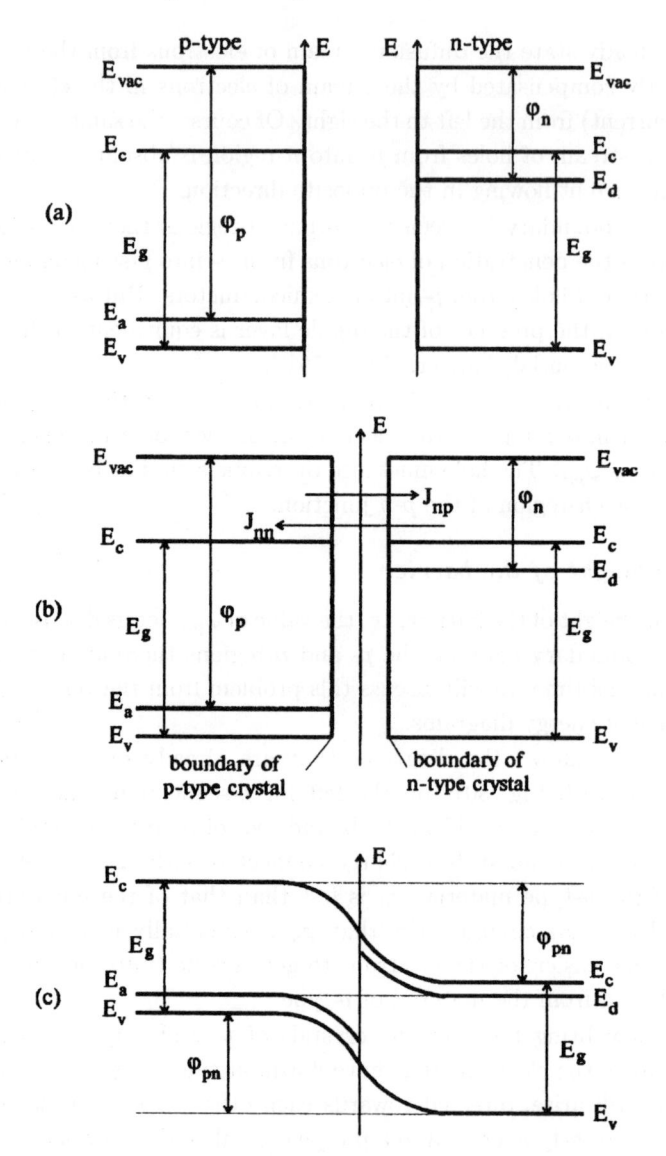

Fig. 48. Formation of the potential barrier of the *p-n* junction and the height of the barrier. (a) the work function of the material of *n*-type φ_n is smaller than that of the *p*-material φ_p; (b) a flow of electrons J_{nn}, directed from an *n*-type crystal to a *p*-type crystal is larger than a flow of electrons J_{np}, from a *p*-type crystal to an *n*-type crystal; (c) in the steady state the energy levels of the electrons in the *n*-type and in the *p*-type parts of the crystal take the same position in the energy diagram. The height of the potential barrier $\varphi_{pn} \approx E_g$.

which both flows will be equal, i.e. a flow of electrons from the p- into the n-region will be equal to a flow from the n- to the p-region.

When is it going to happen?

It will when the energy levels of the electrons in the n-type and in the p-type parts of the crystal takes the same position in the energy diagram.

While the energy level of the electrons in p-region is lower than that in n-region [Fig. 48(b)], it is energetically expedient for the electrons to pass from the n- into the p-region. A stationary state will be established only in the case when the transition neither from n-region into p-region, nor backward will lead to the acquiring or to the loss of energy by the electrons — i.e. when the same value of energy will correspond to the energy levels of the electrons in p- and n-regions.

That steady state is reflected in the energy diagram [Fig. 48(c)], which looks quite familiar to us [compare this to Fig. 47(a)]. It is seen in Fig. 48(c) that the same value of energy corresponds now to the level E_a of the electrons in p-region and to the level E_d of the electrons in n-region, with an energy barrier φ_{pn} in height between the p- and n-region.

Comparing Fig. 48(b) and Fig. 48(c) we find that

$$\varphi_{pn} \approx \varphi_p - \varphi_n \tag{35}$$

In the case considered by us, when in the p-type region electrons are at the shallow acceptor level, and in the n-type region they are at the shallow donor level [Fig. 48(a)], the difference $\varphi_p - \varphi_n$ approximates E_g and consequently $\varphi_{pn} \approx E_g$.

5.2.2 Depletion layer. Width of the barrier

So, between the p- and n-regions in the p-n junction there is an energy barrier, whose height is as follows: for germanium junctions, it is ~ 0.7 eV, for silicon junctions it is 1 eV and for GaAs junctions, it is ~ 1.4 eV. Any electron which, while being in the steady state, would like to penetrate from the n-region into the p-region, or any hole which would like to leave the p-region for the n-region would have to overcome that barrier [Fig. 48(c)].

Since the majority of electrons in the conduction band (and the majority of holes in the valence band) have the energy of the order of kT (i.e. only 0.026 eV at room temperature), it is clear that for the majority of carriers the barrier on the boundary between p- and n-regions seems to be a mountain of a tremendous height, which the carriers are unable to climb at all.

Moreover, the majority of carriers are not only unable to overcome that energy barrier, but they are unable to move even a bit inside the barrier.

Therefore, the region of the energy barrier presents a layer practically depleted of free carriers, electrons and holes. But since it is so, it will be quite simple for us to calculate all the main characteristics of the barrier. Indeed, to the right of the boundary of p-n junction in the n-region, every cubic centimetre of the semiconductor contains N_d positively charged donors. The concentration of electrons and holes in the region of the barrier being negligibly small, as we have established just now, it is clear that the volume charge density ρ in the barrier will be determined by the concentration of the donors: $\rho = qN_d$. In the p-region the volume charge density will be determined by the concentration of the negatively charged acceptors: $\rho = -qN_a$ [Fig. 49(b)].

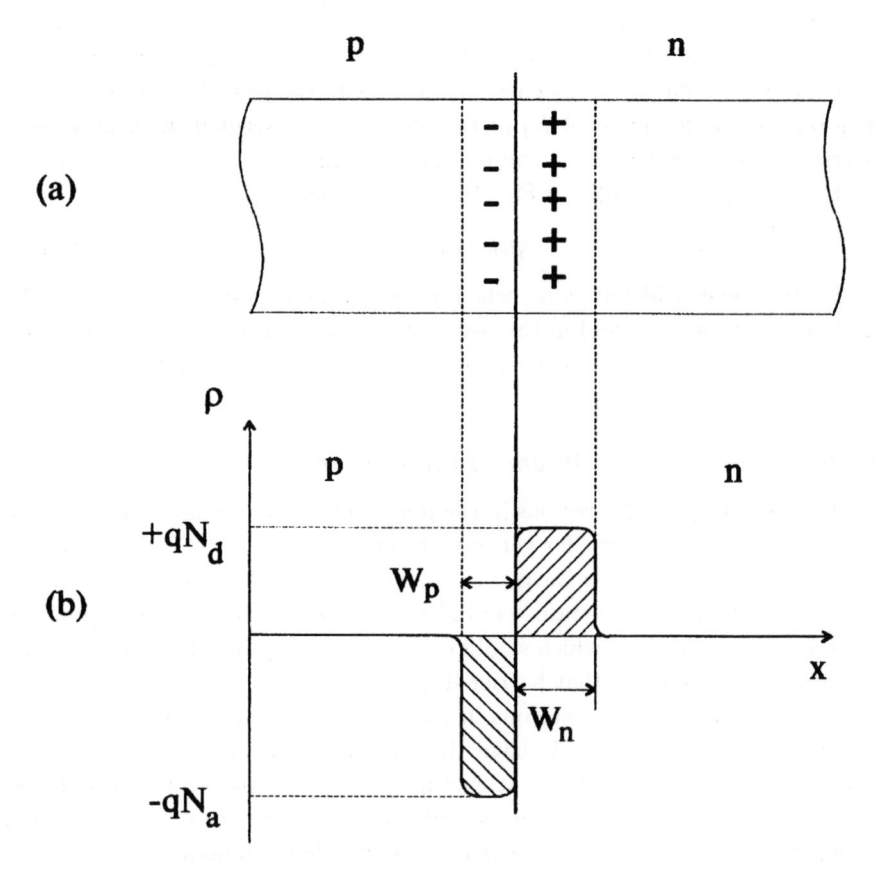

Fig. 49. Space charge distribution into the p-n junction (a) qualitative diagram [(cmp. to Fig. 47(c)]; (b) the dependence $\rho(x)$.

If the volume charge density ρ is a constant, determining the distribution of the field in the barrier and the width of the barrier W is quite a simple problem which has already been solved by us [see Eq. (34) and Fig. 40(b)]. The dependence $F(x)$ in either "half" of the junction will be a straight line. The slope of that line will be determined by the level of doping [Eq. (34), Fig. 40(b)].

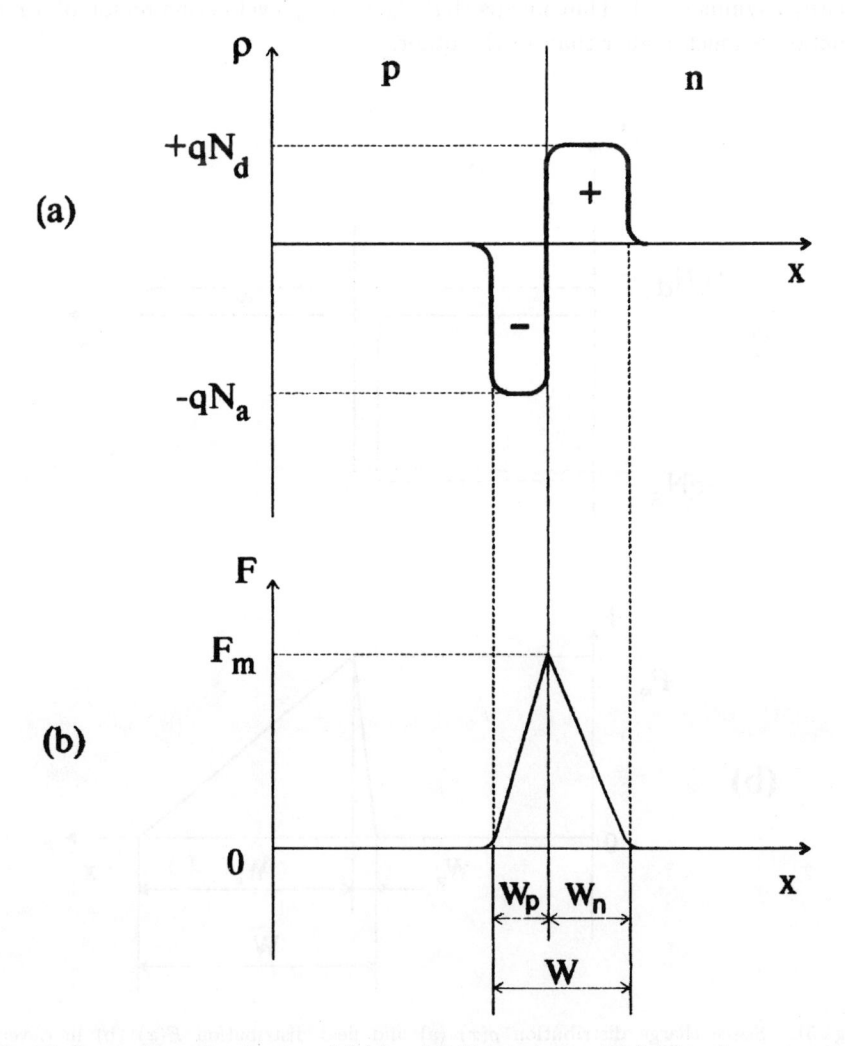

Fig. 50. Space charge distribution $\rho(x)$ (a) and field distribution $F(x)$ (b) in the p-n junction.

The dependence $F(x)$ in the p-n junction is shown in Fig. 50(b). The electric field is maximal in the plane of the p-n junction and decays linearly when moving off the junction. The slope of the dependence $F(x)$ in every region is determined by the level of doping, and the greater the doping impurity concentration, the larger the slope.

The p-n structures that are in practical use in the majority of cases are greatly asymmetrical. That means that the doping level in one region of a p-n junction is much higher than in the other.

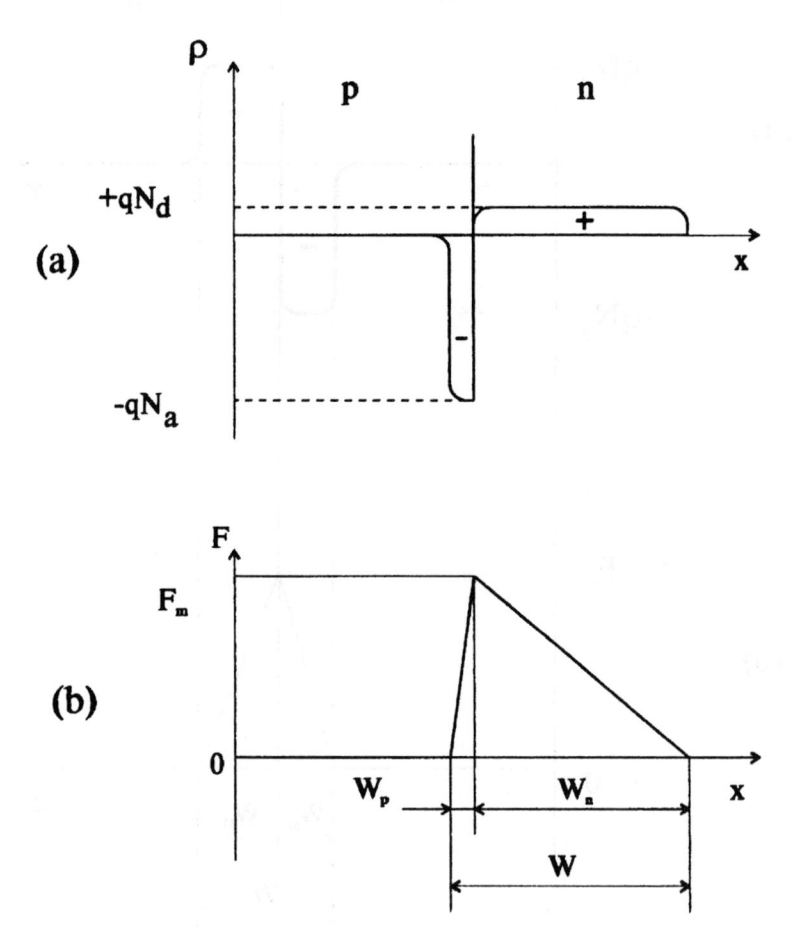

Fig. 51. Space charge distribution $\rho(x)$ (a) and field distribution $F(x)$ (b) in a very asymmetrical p-n junction.

Figure 51 shows the distribution of the field in the p-n junction in which the p-region has been doped much higher than the n-region, i.e. $N_a > N_d$. It is seen in the picture that in this case the width of the space charge region W_p in the p-region is considerably less than W_n, and the distribution of the field $F(x)$ has the shape of a right-angled triangle. Practically all the voltage drop $U_{pn} = 1/2 F_m W_n$ falls on the n-region. The calculation of the barrier parameters in this case does not present any difficulty [compare this with Eqs. (32)]:

$$W \approx W_n = \left(\frac{2\varepsilon\varepsilon_o U_{pn}}{qN_d}\right)^{1/2} \;;\quad E_m \cong \left(\frac{2qN_d U_{pn}}{\varepsilon\varepsilon_0}\right)^{1/2} \qquad (36)$$

The greater the concentration of donors N_d, the larger the field in the barrier and the smaller its width.

In a silicon power high-voltage rectifying diode the donor concentration in the n-region is $N_d = 10^{13}$ cm^{-3}. The acceptor concentration in the p-region is about 10^{18} cm^{-3}, i.e. 100 000 times greater; the junction is really very asymmetrical. Substituting the values $U_{pn} = 1.1$ V, $\varepsilon = 12.5$, $N_d = 10^{19}$ m^{-3} into Eq. (36), we obtain that in such a diode the field in the barrier is $E_m \approx 1.8 \cdot 10^3$ V/cm, and the width of the barrier is $W \approx W_n \approx 12$ μm.

In silicon low voltage diodes, the donor concentration in the n-region may be about 10^{17} cm^{-3}. In such a diode the width of the space charge region is as small as ~ 0.12 μm, while the maximum field in the p-n junction is about $1.8 \cdot 10^5$ V/cm.

5.2.3 Wonderful equilibrium

Let us imagine that there is a tiny piece of a semiconductor from which two metal wires are stretched [Fig. 52(a)]. Let us assume that someone has told us that an acceptor impurity has been introduced into the left part, and a donor impurity into the right part of this piece. Then we know that there is a p-n junction inside the semiconductor. Let us close the outputs of the diode with a very sensitive galvanometer [Fig. 52(b)]. If one can be sure of anything, it is the fact that there won't be any current registering the galvanometer, no matter how sensitive it may be. Indeed, the presence of current would mean the violation of the law of conservation of energy: we would obtain energy (at least that energy which is spent on the Joule heating of the wires), without expending any other energy.

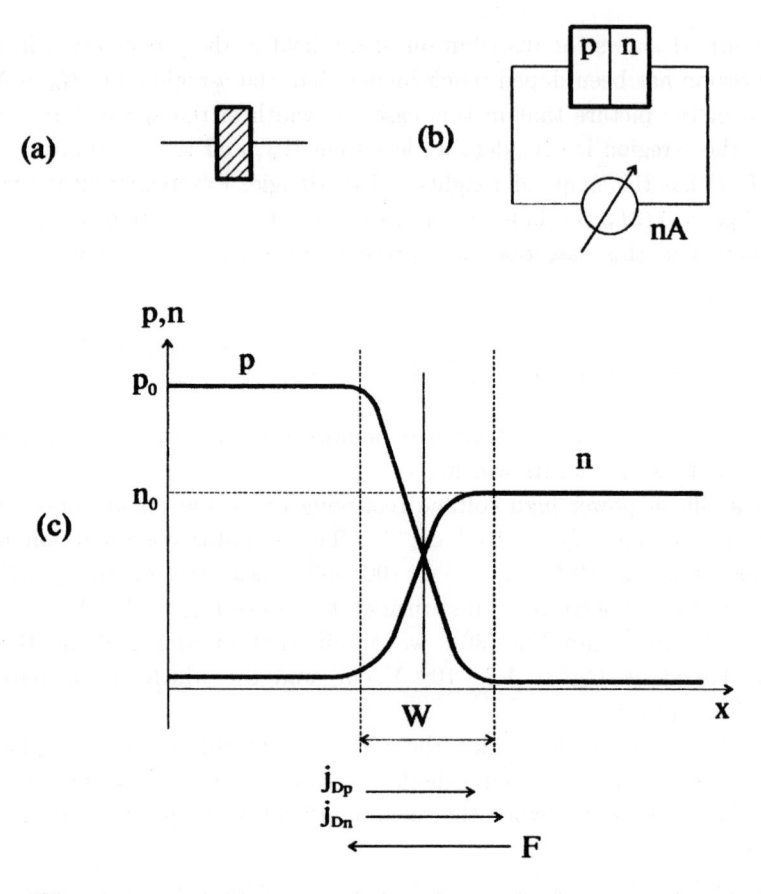

Fig. 52. When no external voltage is applied to the sample, the current through the *p-n* junction is equal to zero in spite of the fact that the derivatives of the concentrations with respect to the coordinate for the electrons dn/dx and for the holes dp/dx are very great. At every point of the barrier the electric field creates a drift current directed towards the diffusion current and compensates it absolutely!

On the other hand, if we recall what we have learnt of the *p-n* junction, such a conclusion may just seem incredible.

Indeed, within the depletion layer at a distance of a few micrometres or even fractions of a micrometre, the free carrier concentration decays very much [Fig. 52(c)]. A simple calculation (we'll do it a bit later) shows that within the narrow depletion layer the carrier concentration decreases millions and even billion of times.

Consequently, the derivatives of the concentrations with respect to the coordinates for the electrons dn/dx and for the holes dp/dx are very great. And consequently, in accordance with Eq. (23), in the region of the barrier, there are strong diffusion currents.[a]

Why then is the current in the circuit equal to zero?

At every point of the barrier, the electric field creates a drift current directed towards the diffusion current and compensates it absolutely!

In order to imagine how large diffusion and drift currents are when they balance each other inside the barrier, we will evaluate, say, the diffusion current density. We could, if we liked, calculate the exact dependencies of the diffusion current density of holes j_{Dp} and of electrons j_{Dn} on the coordinate. But we will restrict ourselves to a more simple task and estimate the values j_{Dp} and j_{Dn} to the order of their value. We will assume that inside the depletion layer, the concentration of holes $p(x)$ and that of electrons $n(x)$ decay linearly, i.e. within the space charge region the values of the derivatives dn/dx and dp/dx are constant.

As an example, we will suppose that we are considering a silicon p-n junction with the concentration of donors in a weakly doped region being 10^{16} cm^{-3}. Let the concentration of acceptors in a high doped region N_a be equal to 10^{18} cm^{-3}. The junction is very asymmetrical, therefore $W \approx W_n$. Using Eq. (35) we can easily establish that $W \approx W_n \approx 0.4$ μm. The mean values of the derivatives dn/dx and dp/dx are

$$\frac{dn}{dx} \sim \frac{n_0}{W} \approx 2.5 \cdot 10^{20} \text{ cm}^{-4},$$

$$\frac{dp}{dx} \sim \frac{p_0}{W} \approx 2.5 \cdot 10^{22} \text{ cm}^{-4}.$$

To calculate the values j_{Dp} and j_{Dn} we will have to calculate the diffusion coefficient values of electrons and holes in silicon: D_n and D_p. Making use of the Einstein relation (22) and the mobility values for the electrons and holes indicated in Table 3, we will find that in silicon at room temperature $D_n = 34$ cm^2/s, $D_p = 13$ cm^2/s. Then by Eq. (23) we will find that in this case $j_{Dn} \approx 1400$ A/cm^2, $j_{Dp} \approx 54000$ A/cm^2. The total density of the diffusion current will be $j_D = j_{Dn} + j_{Dp} \approx 5.5 \cdot 10^4$ A/cm^2. a)

The diffusion current density grows with the growth of the level of doping, because the values n_0 and p_0 increase and the width of the space charge region decreases. For the actual levels of doping $N \sim 10^{17} \div 10^{19}$ cm^{-3} the value j_D can be millions of Ampere/cm^2!

And yet the hand of the galvanometer, able to register a current as small as 1 pA (1 pA = 10^{-12} A) does not deviate at all. The current, created by the field of the barrier, compensates to that precision the diffusion current in every point of the space charge region.

---

[a] Let us recall that though the diffusion flows of electrons and holes in the p-n junction are directed towards each other, the diffusion currents of electrons and holes flow in the same directions, and, consequently, they are added together.

There is one more remarkable kind of equilibrium which should take place so that no current might flow across the p-n junction (without the bias). As we have seen, there is a powerful stream of majority carriers — electrons which try to penetrate from the n-region of a diode into the p-region, but cannot do it, being held up by a strong field of the barrier. However, if we study the band diagram very closely, we can see that there is also a thin spring of minority carriers — electrons in the p-region, the field of the barrier does not hinder, but helps them leave the p-region for the n-region. Indeed, look at Fig. 53. It shows a band diagram, familiar to us [Compare it to Fig. 47(c)]. The only difference is that Fig. 53 shows only the upper half

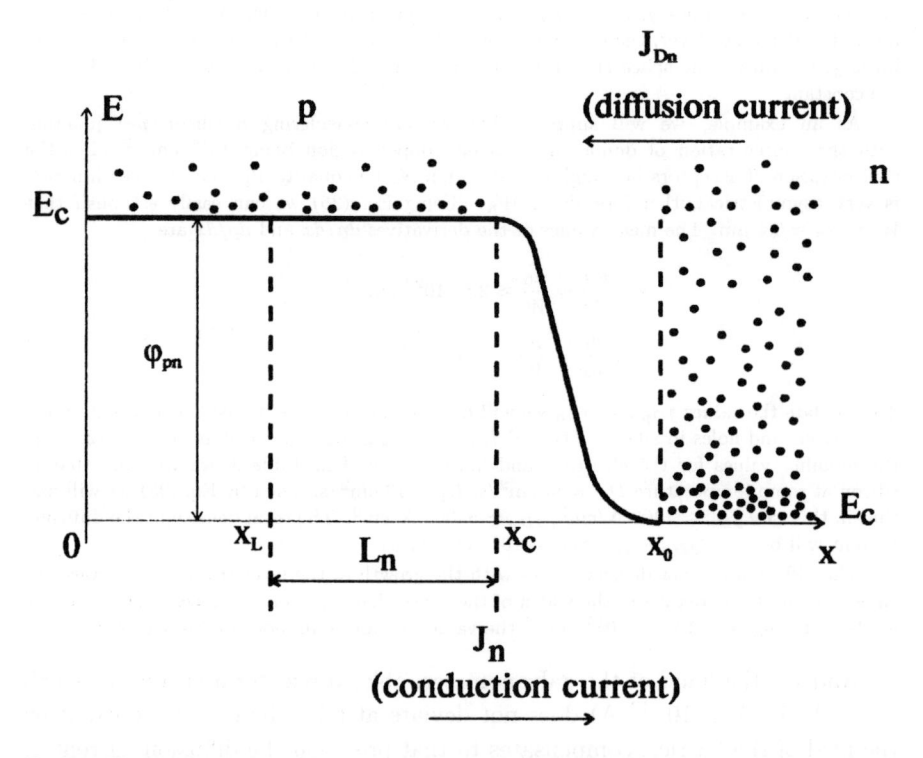

Fig. 53. The profile of the conduction band bottom for the p-n junction. For the electrons in the p-region (minority carriers), the potential barrier of the p-n junction is a "hilloc". Any electron brought on that hilloc, immediately gets down from the p- into the n-region. So there appears the drift electron flow J_n from the p- into the n-region (the lower arrow). In a steady state, this flow is compensated by the diffusion flow of the hot electrons from the n- to the p-region (the upper arrow).

of Fig. 47(c). Figure 53 demonstrates the profile of the conduction band bottom $E_c(x)$ for the p-n junction. The energy in Fig. 53 is measured from the E_c level in a semiconductor of n-type. Points x_0 and x_c correspond to the boundaries of the space charge region. Besides, there is also point x_L in Fig. 53. It is at the distance L_n from the boundary of the space charge region. The symbol L_n designates the diffusion length of electrons in the p-region, equal, as we remember (see Eq. (26)), to the following: $L_n = (D_n \cdot \tau_n)^{1/2}$, where D_n is the electron diffusion coefficient, τ_n is the electron lifetime.

Please note again that for the electrons, the majority carriers in the n-region, the bend of the conduction band bottom E_c corresponds to the potential barrier which must be overcome in order to penetrate into the p-region.

Now what about the minority carriers, electrons in the p-region? It is seen in Fig. 53 that there is a hilloc in the band diagram that corresponds to those electrons. After they are brought on that hilloc, the electrons immediately get down from it into the n-region. (As we have already seen many times [say, in Fig. 47(c)], the field of the barrier is directed in such a way that it tends to push the electrons into the n-region). The same refers of course to holes — the minority carriers in the n-region of a diode [see Fig. 47(d)]. For them the dependence $E_v(x)$ is also a hilloc. After they get on it, the holes flow up into the p-region (The field of the barrier pushes the positively charged holes into the p-region).

Since the concentration of minority carriers is small, the current density, conditioned by the flow of electrons from the p- into the n-region, and by the flow of holes from n- into p-regions, is also small. However, as we will soon see this current plays a very important role in the work of diodes and transistors.

The current we are considering now is called the *saturation current* and is designated by the letter j_s. Why it had been given such a name will become clear a bit later. This current, as we have seen, consists of two components: electrons (from the p- to n-region) and holes (from the n- to p-region).

Now let us determine the electron component of the saturation current density j_{ns}.

But first let us speak of the fate of the electrons existing in the depth of the p-region, to the left of point x_L. Each of those electrons, being born in the conduction band (from the acceptor level or from the valence band [see Fig. 9(a)], lives on the average, a time span of τ_n. During that time it rushes chaotically, participating in thermal motion, and departs from the point of its birth to a distance $\sim L_n = (D_n \cdot \tau_n)^{1/2}$. Then it perishes (recombines).

The larger the electron concentration n_p and the shorter their lifetime τ_n, the greater will be the number of carriers, perishing in a unit time per unit volume n^* : $n^* = n_p/\tau_n$. But since in the stationary state the electron concentration does not change, it is clear that the same number of carriers n^* will be generated per unit time per unit of volume.

And now let us discuss a possible fate of electrons, born in the region between points x_L and x_c. Their fate differs from that of their neighbours born to the left of point x_L only in one aspect. Having been born, and having been wandering chaotically for the time τ_n about the crystal, they may cross the fatal line — appearing to be to the right of point x_c. Should that happen, the electric field of the barrier will instantly catch them and toss them to the n-region (They will come down the potential hilloc to point x_0).

These electrons, born in the p-region at a distance of $x < L_n$ from the space charge region boundary, after they have "crossed the line", make up the electron component of the current density j_{ns}.

In order to estimate the value j_{ns} we will assume that not a single electron born to the left of point x_L will get to the space charge region. And all the electrons, born between points x_L and x_c at the diffusion length distance L_n from the space charge region boundary, will be carried by the barrier field to the n-region.

Then the number of electrons carried per unit area of the barrier from the p-region to the n-region per unit time will be equal to the number of electrons $n^* = n_p/\tau_n$, born per unit volume in a unit time, multiplied by the length L_n. And the charge, carried by them, i.e. the current density j_{ns} will be as follows:

$$j_{ns} = qL_n \frac{n_p}{\tau_n}. \tag{37a}$$

Taking into account that $n_p = n_i^2/N_a$, the equation for the value j_{ns} is often written as

$$j_{ns} = q\frac{n_i^2}{N_a} \cdot \frac{L_n}{\tau_n}, \tag{37b}$$

or, considering that $L_n = (D_n \tau_n)^{1/2}$,

$$j_{ns} = q\frac{n_i^2}{N_a}\left(\frac{D_n}{\tau_n}\right)^{1/2} \tag{37c}$$

Let us first estimate j_{ns} in order of its value. Assuming that $\tau_n = 10^{-6}$ s, $N_a = 10^{16}$ cm^{-3}, and using Tables 2 and 3, we will find that at room temperature for the germanium p-n junction, $j_{ns} \approx 10^{-4}$ A/cm^2; for the silicon p-n junction, $j_{ns} \approx 10^{-11}$ A/cm^2; for gallium arsenide, $j_{ns} \approx 10^{-17}$ A/cm^2.

Those are of course not powerful streams of majority carriers making thousands and tens of thousands of Ampere/cm^2, but even for the gallium arsenide p-n junction, the presence of such a current of minority carriers j_{ns} means that $j_{ns}/q \approx 10^2$ electrons pass every second across every square centimetre of the p-n junction from the p- region to the n-region. And for the germanium p-n junction, it is $\sim 10^{15}$ electrons.

Meanwhile the conditions of the wonderful equilibrium absolutely demand that should the electron leave the p-region for the n-region, it will be compensated by the transfer of the electron from the n-region to the p-region.

Where from do the electrons appear, those that get in a stationary state from the n-region into the p-region?

As we have already made sure, apart from a tremendous majority of electrons whose energy is $\sim kT$, there are also very energetic, hot electrons in the n-region [See Eq. (12)]. Among them there are also such whose energy E is larger than the height of the energy barrier φ_{pn}. The concentration of such electrons is, of course, very small, $(n \approx n_0 \cdot \exp(-\varphi_{pn}/kT))$, but the concentration of the minority carriers, forming the saturation current, is also very small. The electrons whose energy is larger than the barrier height φ_{pn} cannot be held up by the barrier field in the n-region. These energetic electrons diffuse quite freely from the n-region to the p-region, and the diffusion current of those hot carriers brings as many electrons in the steady state into the p-region from the n-region as the current j_{ns} takes away.

The hole component j_{ps} of this current can be established in exactly the same way as the electron component j_{ns} of the saturation current has been established by us.

$$j_{ps} = qL_p \frac{p_n}{\tau_p} = q \frac{n_i^2}{N_d} \frac{L_p}{\tau_p} = q \frac{n_i^2}{N_d} \left(\frac{D_p}{\tau_p}\right)^{1/2} \tag{38}$$

The values of lifetime of holes τ_p in the n-region, of the hole diffusion coefficient D_p and hole diffusion length $L_p = (D_p \tau_p)^{1/2}$ can of course differ from the corresponding values of the electrons in the p-region.

The full value of the saturation current density j_s is equal to

$$j_s = j_{ns} + j_{ps} = qn_i^2 \left(\frac{1}{N_a}\sqrt{\frac{D_n}{\tau_n}} + \frac{1}{N_d}\sqrt{\frac{D_p}{\tau_p}}\right) \tag{39}$$

If, as it usually happens, the junction is quite asymmetrical, then one of the components of the saturation current can be neglected. For instance, if the p-region of the junction is high doped, so that $N_a \gg N_d$, then, as is seen in Eq. (38), $j_{ps} \gg j_{ns}$. The main contribution into the saturation current is made by holes.

The value j_s is proportional to the square of the intrinsic concentration of carriers n_i^2, therefore it depends very much on the temperature. It follows from Eq. (6) that $j_s \sim n_i^2 \sim \exp(-E_g/kT)$. Knowing the value E_g (see Table 1), it is easy to calculate that if at room temperature (300 K) the value j_s for the germanium diode is, say, ~ 1 mA/cm^2, then at a temperature $+80°$C (353 K), it will make ~ 60 mA/cm^2, i.e., increase 60 times. The current density j_s of the silicon diode with the growth of temperature from 300 to 353 K increases approximately 600 times, and in the gallium arsenide diode, 3500 times.

The temperature $+80°$C corresponds to the maximum possible operational temperature for germanium diodes. The silicon p-n structures can operate until the temperature reaches $150°$C (423 K). At that temperature the saturation current density in silicon diodes increases compared to room temperature ~ 250000 times. (And yet it remains less than the value j_s in the germanium p-n structures at room temperature). The gallium-arsenide diodes can operate until the temperature reaches $250°$C (523 K). Then the value j_s increases, compared to its value at room temperature $\sim 10^{10}$ times! And makes as little as $\sim 5 \cdot 10^{-7}$ A/cm^2. Finally, there are p-n junctions based on silicon carbide, which can operate until the temperature reaches $\sim 600°$C! Such diodes can work quite safely in a bath with melted lead. At temperatures close to ~ 800 K, the saturation current increases in SiC diodes $\sim 10^{30}$ times in relation to room temperature. And yet (make the calculation yourself, assuming that $n_i = 10^{-5}$ cm^{-3}, $N_a = 10^{17}$ cm^{-3}, $D_n = 8$ cm^2/s, $\tau_n = 10^{-8}$ s) the value j_s for SiC diode at 800 K makes only $\sim 5 \cdot 10^{-12}$ A/cm^2.

The greater the width of the forbidden band E_g material, of which the diode is made, the higher is its operational temperature and the smaller the value j_s.

5.2.4 The reverse bias

Having admired the perfection with which streams of electrons and holes are compensated in the p-n junctions, let us violate the wonderful balance. To do it, it is sufficient to apply the voltage, or as it is often said, the bias U_o from the outer source (Fig. 54) to a diode containing a p-n junction. If the "plus" of the source is connected to the n-region of the diode, and the "minus", to the p-region, then it is usually said that a reverse voltage (or reverse bias) has been applied to the p-n junction. That is just the case depicted in Fig. 54.

A diode, as we know, contains three regions: the p-region, the n-region and the space charge region (the barrier) which is between them. In reference to the external source U_0, all the three regions are connected in series. And the

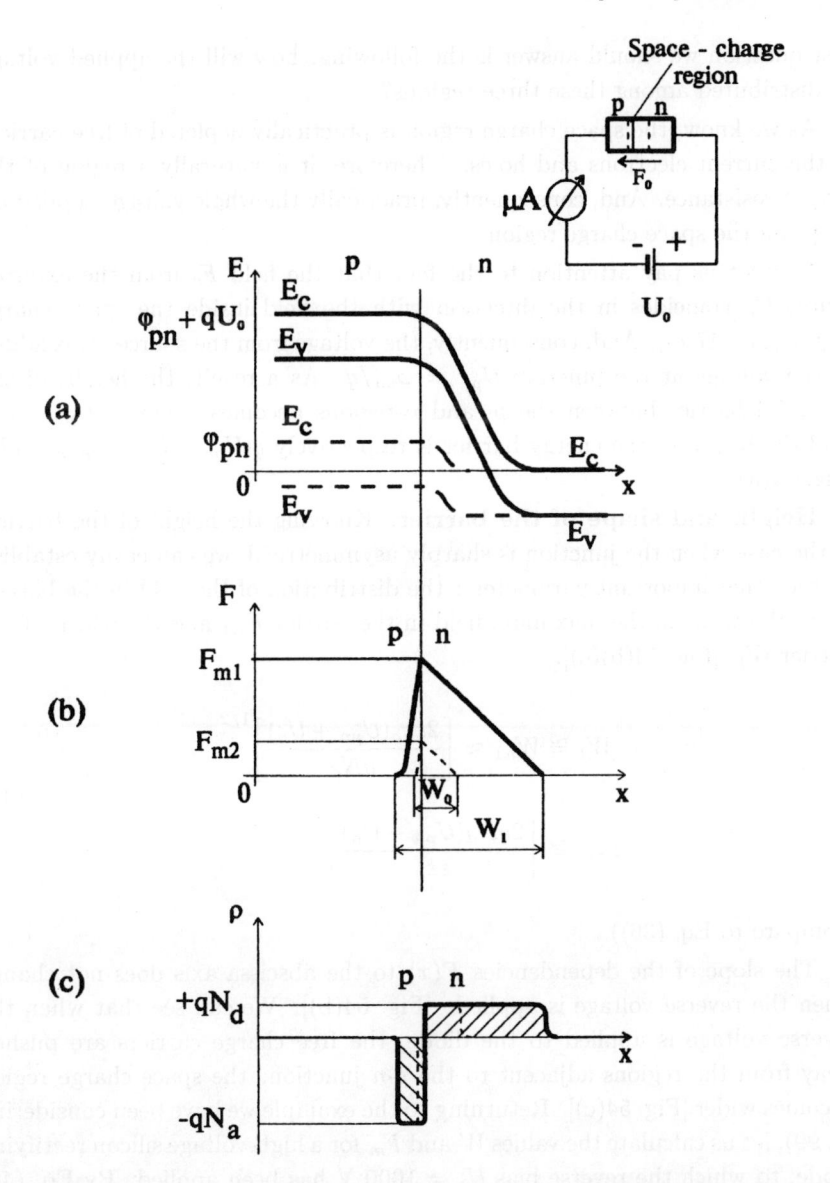

Fig. 54. Band diagram (a), field distribution (b) and space charge distribution (c) for the p-n junction. Dashed curves represent the case when the applied voltage is equal to zero. Solid curves represent the case of the reverse bias. Under the reverse bias the height of the potential barrier increases. The height of the energy barrier is now equal to $\varphi_{pn} + qU_o$. The width of the space charge region and the maximum field F_m increase as well.

first question we should answer is the following: how will the applied voltage be distributed among these three regions?

As we know, the space charge region is practically depleted of free carriers of the current electrons and holes. Therefore, it is naturally a region of the largest resistance. And, consequently, practically the whole voltage applied U_0 drops on the space charge region.

Now let us pay attention to the fact that the field F_o from the external source U_o coincides in the direction with the field inside the space charge region [Fig. 47(c)]. And, consequently, the voltage from the source U_o is added to the voltage at the junction $U_{pn} = \varphi_{pn}/q$. As a result, the height of the potential barrier between the p- and n-regions becomes equal to $U_{pn} + U_o$. And the height of the energy barrier is respectively $q(U_{pn} + U_o) = \varphi_{pn} + qU_o$ [Fig. 54(a)].

Height and shape of the barrier. Knowing the height of the barrier, in the case when the junction is sharply asymmetrical, we can easily establish all the other important parameters: the distribution of the field in the barrier $F(x)$, the value of the maximum field in the barrier F_{m1} and the width of the barrier W_1, [Fig. 54(b)(c)].

$$W_1 \cong W_{n1} \approx \left[\frac{2\varepsilon\varepsilon_o(U_{pn} + U_o)}{qN_d}\right]^{1/2}$$

$$F_{m1} \cong \left[\frac{2qN_d(U_{pn} + U_o)}{\varepsilon\varepsilon_o}\right]^{1/2}$$

(40)

(compare to Eq. (36)).

The slope of the dependencies $F(x)$ to the abscissa axis does not change when the reverse voltage is applied. [Fig. 54(b)]. We can see that when the reverse voltage is applied to the diode, the free charge carriers are pushed away from the regions adjacent to the p-n junction: the space charge region becomes wider [Fig. 54(c)]. Returning to the example we have been considering (p. 99), let us calculate the values W and F_m for a high-voltage silicon rectifying diode, to which the reverse bias $U_o = 1000$ V has been applied. By Eq. (40) we find $F_{m1} = 54$ kV/cm, $W_1 = W_{n1} = 370$ μm. With the increase of the reverse voltage applied, the maximum field F_m increases. Meanwhile for every semiconductor material there is a restricting value of the electric field F_i, which cannot be exceeded "with impunity".

In the field F_i, which for Si and GaAs is $\sim 3 \cdot 10^5$ V/cm, electrons and holes acquire so much energy that they become able, when colliding with the atoms of the crystal lattice, to generate new electrons and holes. This phenomenon, resembling the ionization of gases, is called *impact ionization*. The phenomenon of impact ionization restricts the value of the reverse maximum voltage which can be applied to a diode.

Let us pay attention to the following very important circumstance. Though the value F_i is practically a constant for the given semiconductor material, the reverse maximum voltage U_i which can be applied to the *p-n* junctions, manufactured of one and the same material, may differ thousands of times.

Indeed, let us assume that the reverse maximum voltage U_i which can be applied to a diode, is defined by the condition $F_m = F_i$. Thus defined, this voltage is called the *breakdown voltage*. From Eq. (40) it follows that

$$U_i \cong \frac{\varepsilon\varepsilon_o F_i^2}{2qN_d}. \tag{41}$$

If $N_d \approx 10^{17}$ cm^{-3}, then $U_i \approx 3$ V, if $N_d \approx 10^{14}$ cm^{-3}, then $U_i \approx 3000$ V.

Reverse current. Figure 55 indicates a typical current-voltage characteristic of a reverse-biased germanium *p-n* junction, i.e. the dependence of the current I flowing across the junction on the value of applied reverse bias U_0.

The first thing that strikes you at first glance is this current-voltage characteristic which has nothing in common with the current-voltage characteristic of an ordinary active resistor R. For the resistor, the current is proportional to the applied voltage; the current-voltage characteristic presents a straight line. The current-voltage characteristic of the reverse-biased *p-n* junction, indicated in Fig. 55, is characterised by the saturation of the current. At the reverse voltage, as small as of the order of decimal fractions of volt, the current stops being dependent on voltage. What kind of current is it then?

That is a saturation current I_s which is quite familiar to us.

When no external voltage was applied to the diode, the saturation current was balanced by a stream of majority hot carriers. But now that a reverse bias has been applied to the diode and the height of the barrier has increased, those carriers are no longer able to overcome that barrier. It is enough just to cast a look at Fig. 54(a) to be certain of it.

The carriers whose energy exceeded the height of the barrier, represented by a dashed line, are certainly unable to overcome the barrier represented by

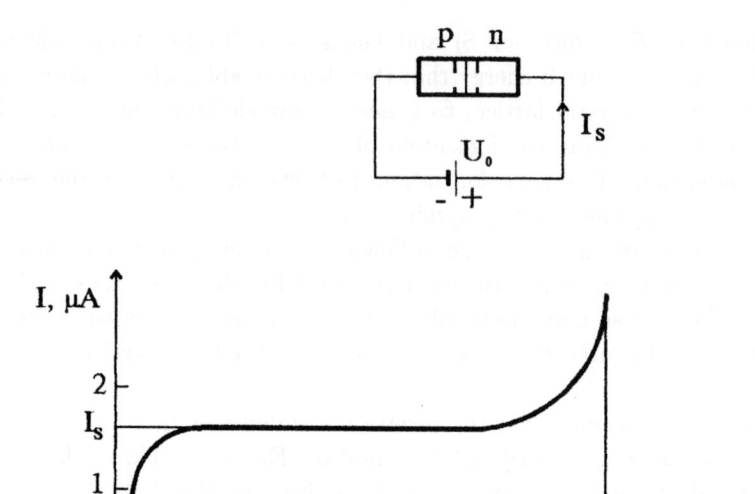

Fig. 55. Current-voltage characteristic of the reverse biased Ge p-n junction.

a solid line. But the applied reverse voltage does not affect the saturation current, i.e. the stream of minority carriers. Exactly the same number of electrons as before will be generated at the distance L_n from the junction and will pass from the p-region to the n-region, forming the electron component of the saturation current I_{ns}. And the same number of holes will pass from the n-region to the p-region forming the hole component of the saturation current I_{ps}. The components of the current, flowing in the opposite direction, those due to the hot carriers, are suppressed due to the grown height of the barrier, and the components of the saturation current do not change. And they display themselves. They flow across the reverse biased p-n junction and form the current $I_s = I_{ns} + I_{ps}$ (Fig. 55), which does not depend on the voltage applied.

Since $I_s \sim n_i^2$ [Eq. (39)], the reverse current grows abruptly with the increase of temperature.

In Fig. 55, it can be seen that at the large reverse bias U_0 the current begins to grow abruptly with the increase of voltage. This effect is caused by the phenomenon we have already mentioned — the impact ionization in the p-n junction. The current-voltage characteristics of the p-n junction at the reverse bias are not always like the dependencies $I(U_0)$ shown in Fig. 55. The dependencies shown in Fig. 56 are much more typical for the silicon and gallium arsenide p-n junctions. The values of the reverse current, measured experimentally, prove to be much larger (tens, hundreds and sometimes millions of times) than the value I_s calculated by Eq. (39). With the growth of the reverse bias, the current does not get saturated, but increases monotonously, proportionally to $U_0^{1/2}$. What does this kind of current-voltage characteristic indicate?

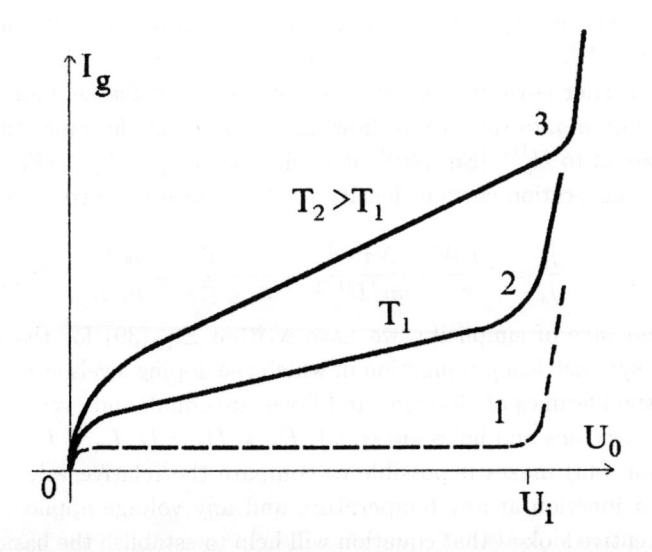

Fig. 56. The qualitative current-voltage characteristics which are typical for the silicon and gallium arsenide p-n junctions. Dashed curve 1 represents the voltage dependence of the saturation current $I_s(U_0)$. Solid curves 2, 3 represent the current, generated in the space-charge region (generation current).

The point is that apart from the saturation current there is one more component of the reverse current. In germanium diodes, it is small compared to the saturation current. But it plays a certain role in silicon and gallium arsenide diodes. The name of this component is "the current, generated in the space-charge region". Or, simply: "the generation current".

There is a strong electric field in the space-charge region. It is clear therefore that any carrier, an electron or hole, born in the space charge region, will be immediately caught up by the electric field and be thrown away: electrons, to the n-region; holes, to the p-region. These carriers, generated in the space-charge region, carried out by the field, form the generation current I_g (Fig. 56). The rate of generation, and consequently the generated current, will be greatest when the deep centre, across which the generation of electrons and holes goes, lies in the middle of the forbidden band (Fig. 12).

In this case, practically the most important one, the generation current density j_g is described by the following equation:

$$j_g = q\frac{n_i}{\tau}W \qquad (42)$$

It is quite clear why the value j_g is proportional to the width of the space charge region W: the greater the value of W, the wider will be the region where any carrier born by generation is caught up by the field and makes its own contribution into the current flowing across the diode. Since the value W is proportional to $U_0^{1/2}$ [Eq. (40)], it is clear that $j_g \sim U_0^{1/2}$ (Fig. 56). The ratio of the generation current density to the saturation current density is as follows:

$$\frac{j_g}{j_s} = \frac{qn_iW}{\tau} \cdot \frac{N\tau^{1/2}}{qn_i^2 D^{1/2}} = \frac{N}{n_i}\frac{W}{\sqrt{D\tau}} = \frac{N}{n_i}\frac{W}{L} \qquad (43)$$

(For the sake of simplicity, we have written Eq. (39) for the case of an absolutely symmetrical p-n junction in which the doping levels are equal: $N_d = N_a = N$, the lifetimes of electrons and holes are equal, and even the diffusion lengths of electrons and holes are equal: $D_n = D_p = D$, $L_n = L_p = L$).

Equation (43) makes it possible to compare the relative values j_g and j_s for any p-n junction at any temperature and any voltage applied. However, just an attentive look at that equation will help to establish the basic property, characterising the ratio j_g/j_s. *With the growth in width of the semiconductor forbidden band E_g the role of the generation current increases very abruptly (exponentially).* Indeed, with the growth of E_g the value of the intrinsic concentration of carriers n_i, decreases very abruptly (see Table 1). That is why the contribution of the generation current to the reverse current in silicon, and especially in galium-arsenide p-n junctions is much greater than in germanium p-n junctions.

Barrier capacity. Let us continue our study of the remarkable properties of p-n junctions. Let us discuss what will define the resistance of the p-n junction to which, apart from the reverse voltage U_0, a small alternating voltage is applied.

The answer to this question may seem somewhat unexpected. With regard to the alternating voltage the reverse-biased *p-n* junction will behave like a capacitor.

The capacitive reactance of the *p-n* junction just like that of any other capacitor, is defined, as is known from any school text book, by the following expression: $X_c = 1/\omega \cdot C$, where $\omega = 2\pi f$ is the angular frequency, and C is the capacitance of the *p-n* junction.

Where does the capacitance of the reverse-biased junction come from? What is its nature and what is it equal to?

The last question is the easiest. The capacitance of the *p-n* junction should be defined by the equation of capacitance for the plane capacitor.

$$C = \frac{\varepsilon \varepsilon_o S}{W}, \qquad (44)$$

where W is the width of the space charge region, S is the area of the junction.

Being able to define the width of the space charge region [Eq. (40)], we can define the capacitance of the junction at any applied voltage and any level of doping.

What is the nature of this capacitance?

Let us recall why an ordinary capacitor — a dielectric plate, supplied with metal coating and presenting an infinitely large resistance for the direct current, can conduct an alternating current. The reason for it is that when an alternating voltage is applied to a capacitor, the latter, during a certain part of the period stores the electric charge, and then, in the other part of the period, gives it away to the external circuit. The periodic charging and discharging of the capacitor maintains the alternating current in the circuit.

And now let us consider the reverse-biased *p-n* junction, to which, besides the constant reverse voltage U_0 a small alternating voltage with the amplitude ΔU_0 is applied (Fig. 57). Solid curves in Fig. 57 correspond to the distribution of the space charge ρ in the *p-n* junction at the reverse bias U_0 (compare to Fig. 49). The dashed curve shows the distribution of the space charge ρ when the voltage $U_0 + \Delta U_0$ is applied to the diode. The voltage at the diode become larger, the space charge region increasing. The free carriers — electrons in the *n*-region and holes in the *p*-region — were pushed out into the external circuit.

The dotted curve indicates the space charge distribution when a voltage $U_0 - \Delta U_0$ is applied to the diode. The voltage at the diode became smaller — the width of the space charge region decreases. And that means that the free

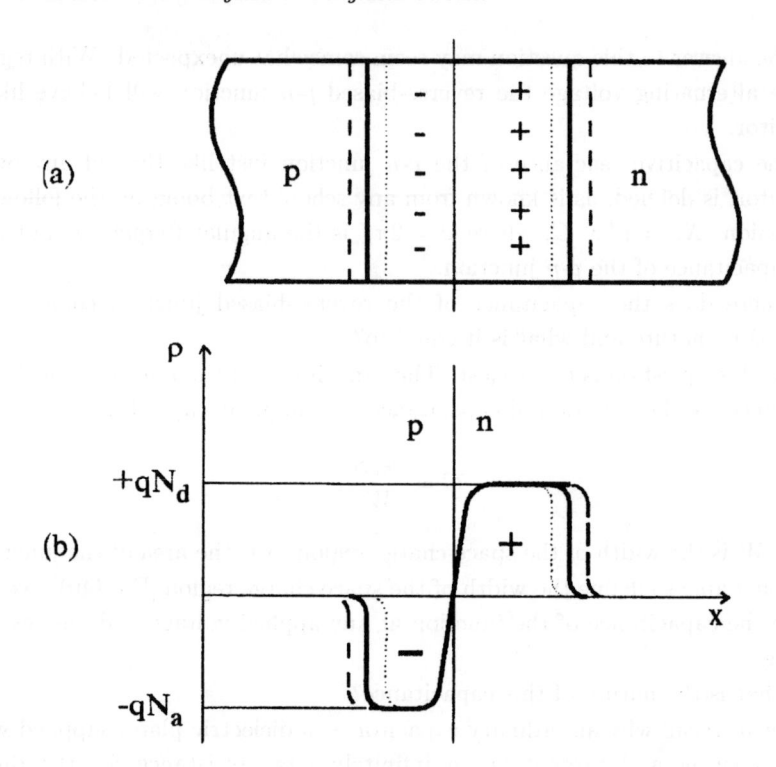

Fig. 57. Distribution of the space charge ρ in the p-n junction at the reverse bias (compare to Fig. 49). Solid curves represent the case when a reverse bias U_0 is applied to the diode. Dashed curves represent the case when the reverse bias is equal to $U_0 + \Delta U_0$. Dotted curves represent the case when the reverse bias is equal to $U_0 - \Delta U_0$.

carriers had been "called" from the external circuit, and filled up those parts of the diode where there were no carriers at U_0 voltage.

Thus, it is clear now that the p-n junction is capable of storing free charge carriers in one part of the period, giving them away into the external circuit in the other part of the period, i.e. the p-n junction possesses capacitance. That type of capacitance of the p-n junction is called the *barrier capacitance*. Mark that as it follows from Eq. (44), the value of the barrier capacitance depends on voltage. The greater the applied voltage U_0 the greater the value W [Eq. (40)] and the smaller the value of the capacitance C. Capacitors whose capacitance depends on voltage are called nonlinear capacitors. So, the p-n junction is a nonlinear capacitance.

Sometimes the capacitive current, caused by the barrier capacitance, prevents the diode from high frequency operation. In this case it is natural to make the value C as small as possible. The shortest way to achieve it is to decrease the operational area of the diode. That is why the area of diodes, meant for operating in microwave circuits, is $10^{-4} - 10^{-6}$ cm^2. The value of the barrier capacitance of such p-n structures is equal to decimal fractions and sometimes to hundredth of picofarade.

However, there are such diodes whose action is based directly on the properties of the barrier capacitance of the junction.

5.2.5 The forward bias

Figure 58 shows a diode with a p-n junction to which a forward bias is applied. The "plus" of the source is now connected to the p-region of the diode, the "minus" — to the n-region. Now the field F_0 from the external voltage source U_0 is directed towards the field, existing within the space charge region [Fig. 47(c)]. And, consequently, the voltage drop of the external source U_0 is *subtracted* from the barrier potential $U_{pn} = \varphi_{pn}/q$, which had existed before the external bias was applied. As a result, the height of the potential barrier between the p- and n-regions becomes equal to $U_{pn} - U_0$. And the height of the energy barrier is accordingly

$$q(U_{pn} - U_0) = \varphi_{pn} - qU_0 \tag{45}$$

The height of the barrier. Pay attention to the fact that if Eq. (45) is valid, then at very small voltages ($U_0 \approx 0.7$ V for the germanium junction, $U_0 \approx 1.1$ V for the silicon junction and $U_0 \approx 1.4$ V for the gallium arsenide junction), the barrier disappears. Its height becomes equal to zero. And what will happen if a larger forward bias is applied? Will the voltage at the barrier change its sign? Will it become negative? What does it mean physically?

Those are very serious and interesting questions. The answer to them will be given later, when we analyze the properties of the forward biased junction. Now, to begin with, we will assume that when the forward bias applied, U_0, is smaller than the height of the barrier U_{pn}.

Forward Current. Since the height of the potential barrier is lowered at forward bias, the electrons, no longer held up by the barrier field, are injected[b] into the p-region, the holes are injected into the n-region of the diode (Fig. 59).

---

[b]The word "inject" comes from the Latin *injicio* — "throw in."

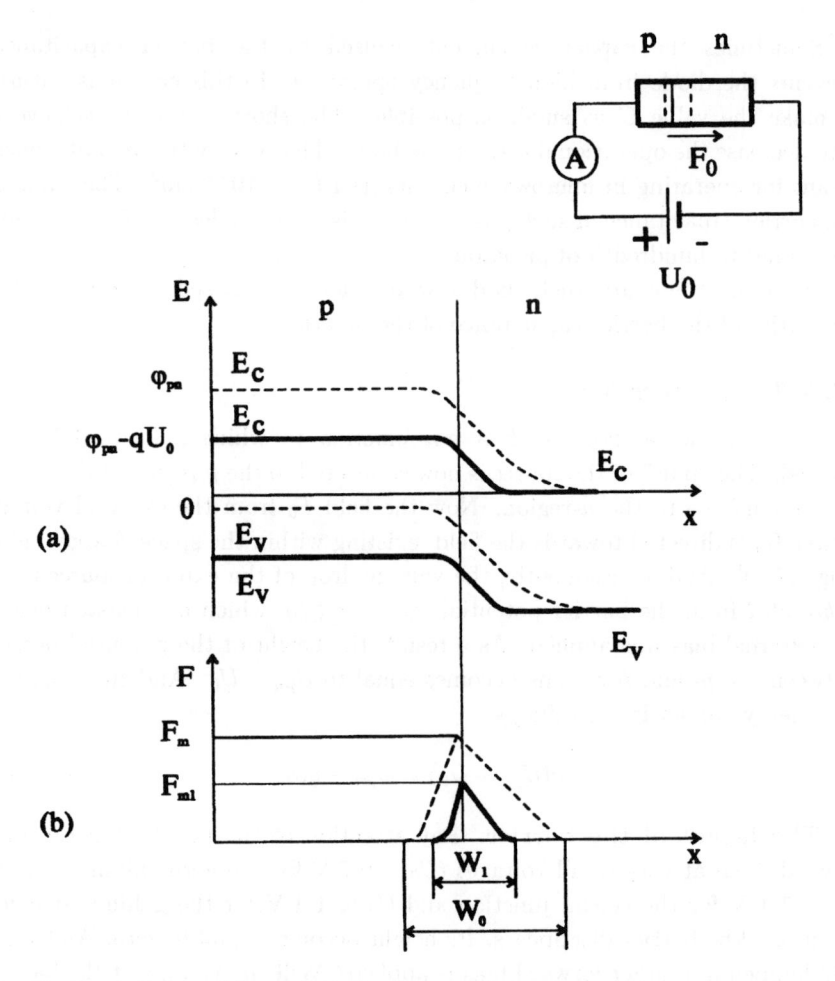

Fig. 58. Band diagram (a) and field distribution (b) for the p-n junction. Dashed curves represent the case when the applied voltage is equal to zero. Solid curves represent the case of the forward bias.

In Fig. 59, a dashed line indicates the picture exactly repeating Fig. 53. While there is no bias, only those electrons whose energy is greater than the height of the barrier φ_{pn} "break through" from the n- to the p-region. Their concentration is $n \approx n_0 \exp(-\varphi_{pn}/kT)$. The diffusion current J_n formed by them is exactly equal to the electron component of the saturation current J_{ns} and balances it.

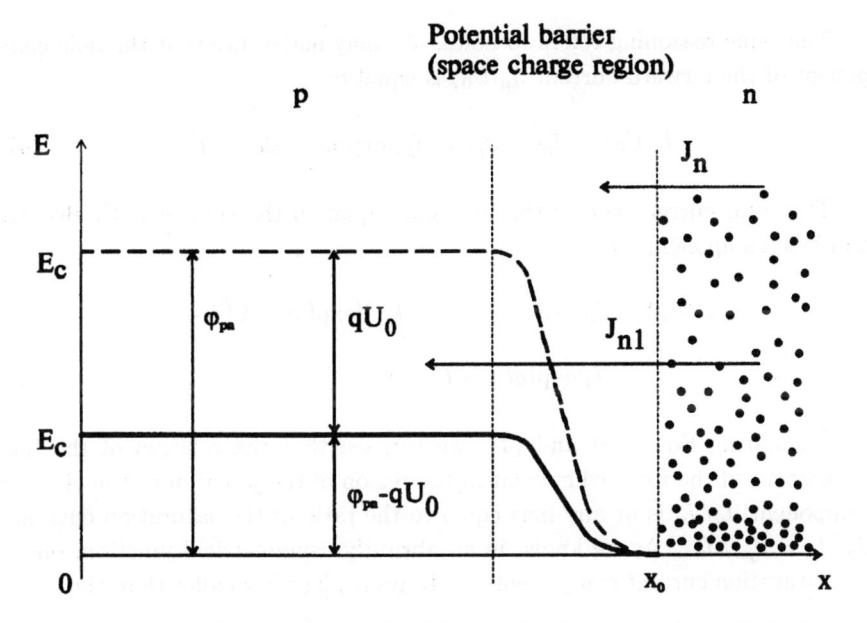

Fig. 59. The profile of the conduction band bottom for the p-n junction. Dashed curves represent the energy barrier profile when the applied voltage is equal to zero. The solid curve is the energy barrier profile at the forward bias U_0. The barrier height became smaller by the value qU_0. All those electrons whose energy exceeds the height of the new barrier $(\varphi_{pn} - qU_0)$ are able to break through to the p-region.

A solid curve $E(x)$ corresponds to a forward bias U_0. The barrier height became smaller by the value qU_0. Now all those electrons whose energy exceeds the height of the new barrier $(\varphi_{pn} - qU_0)$ are able to break through to the p-region. The concentration of such electrons is $n_1 \approx N_0 \exp[-(\varphi_{pn} - qU_0)/kT]$. Thus, the number of electrons, capable of overcoming the barrier which has become lower, has increased $n_1/n = \exp(qU_0/kT)$ times. The diffusion current, formed by them, has increased the same number of times and now is $J_{n1} = J_{ns} \exp(qU_0/kT)$.

The saturation current is not affected by forward bias (or by reverse bias either). The electron current from the p-region into the n-region (the current of minority carriers) is still equal to J_{ns} and flows toward the diffusion current J_{n1}.

Thus, the electron component of the forward current $J_n(U_0)$ will be equal to

$$J_n(U_0) = J_{n1} - J_{ns} = J_{ns}[\exp(qU_0/kT) - 1] \qquad (46)$$

The same reasoning refers to holes. We may make sure that the hole component of the forward current $J_p(U_0)$ is equal to

$$J_p(U_0) = J_{p1} - J_{ps} = J_{ps}[\exp(qU_0/kT) - 1] \qquad (47)$$

The total current across the junction, equal to the sum of both electron and hole components, is

$$J(U_0) = J_p + J_n = (J_{ps} + J_{ns})[\exp(qU_0/kT) - 1]$$
$$= J_s[\exp(qU_0/kT) - 1]. \qquad (48)$$

Comparing Eqs. (46) and (47) we can see that the relation of the hole component of the forward current in the region of the junction to the electron component J_p/J_n is at any bias equal to the ratio of the saturation currents: $J_p/J_n = J_{ps}/J_{ns}$. As we know, in an abruptly asymmetrical junction, one of the saturation current components is always negligibly smaller than the other.

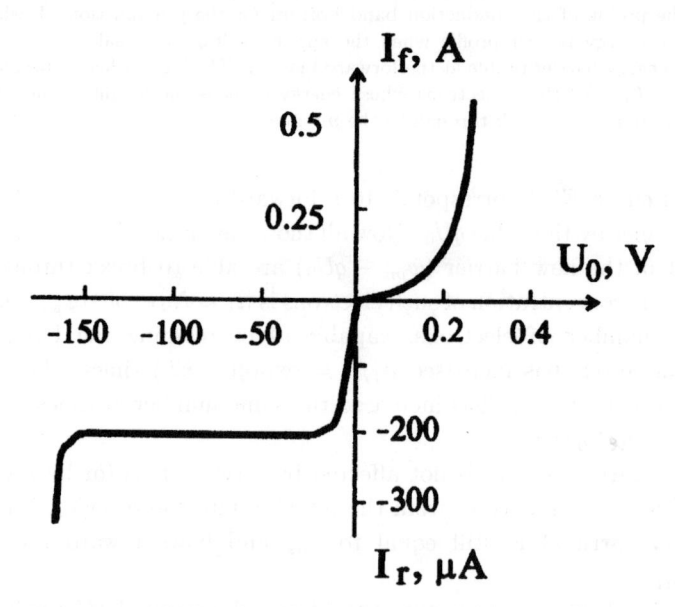

Fig. 60. Current-voltage characteristic of a Ge p-n junction. Positive values of U_0 correspond to the forward bias. Pay attention to the different values of the forward (positive) and the reverse (negative) voltage and the current.

For instance, if the p-region of the junction is strongly doped, then $j_{ps} \gg j_{ns}$, and consequently, at any bias, the main contribution into the common current in such a junction is made by holes. Equation (48) describes the dependence of the current across the p-n junction on voltage at any bias — forward and reverse. Only with the forward applied voltage the value U_o should be considered positive, and with the reverse voltage, negative. With the forward bias the forward current density grows abruptly, exponentially with the applied voltage. And with the reverse bias it saturates at the level $J = -J_s$ at the reverse bias U_0 equal to several kT/q (Fig. 60).

Pay attention to a very important circumstance. When deducing Eq. (48) we thought, as before, that the whole voltage U_0, applied to the diode, drops at the p-n junction, i.e. that the resistance of the p-n junction is much larger than all the other resistance connected in series with it: the resistance of the n-region, of the p-region, the resistance of contacts to the diode, etc.

At the reverse voltage, that supposition was quite justified. The resistance of the barrier (of the p-n junction) was great even when the bias was equal to zero and it grew even more with the growth of the reverse voltage. It is quite different in the case of a forward bias. With the growth of the forward voltage the resistance of the barrier swiftly decreases! When the forward bias is large enough, the resistance of the barrier becomes so small that an appreciable part of the bias, applied to the diode, begins to drop not on the p-n junction, but on the contacts and on the resistance of the p- and n-regions. As a result, the situation of applying a forward voltage, exceeding the initial height of the barrier U_{pn}, is absolutely impossible. The utmost that can be done by increasing the forward voltage is to diminish very much the height of the barrier, making the voltage drop at the barrier very small. In that case the density of the current across the diode is, as a rule, very great. The concentration of carriers in the region "where there used to be a barrier before" is also very great. At such forward bias, approximating the initial height of the barrier U_{pn}, it makes no sense any longer to speak of the voltage drop at the barrier.

Injection. We have just established that if a forward bias is applied to a diode, the height of the potential barrier decreases and the electrons from the n-region are injected into the p-region of the diode. And the holes from the p-region are injected into the n-region. It is interesting, useful and instructive to trace the fate of these injected carriers. Let us begin, say, with the holes injected into the n-region. When there, they immediately get into the position of unwanted guests, the excess nonequillibrium minority carriers, and begin to die out — to recombine, their typical lifetime being τ_p. Recombining, they will move under diffusion from the p-n junction region, where most of them are, into the depth of the n-region. If the forward current flows across the junction for a long time, then there is a stationary distribution of holes in the n-region (Fig. 61). The concentration of holes is maximum at the p-n junction from where they are injected, and it decays into the depth of the n-region. A typical distance at which the holes penetrate into the n-region is equal to the diffusion length of holes $L_p = (D_p \cdot \tau_p)^{1/2}$. The picture resembles the diffusion of nonequillibrium carriers discussed in Chapter 2. But there the excess carriers were created by light, and here by injection across the p-n junction (compare this to Fig. 25).

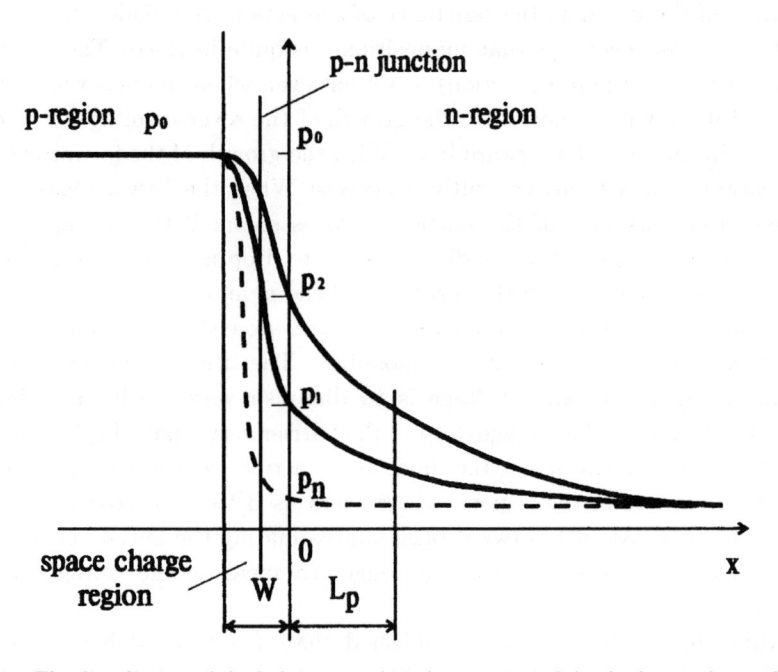

Fig. 61. The distribution of the holes injected in the n-region of the diode at a forward bias. The dashed curve represents the case when the current through the p-n junction is zero. Solid curves represent two different values of the forward current. The greater the forward current, the greater the concentration of holes that are being injected from the p-region into the n-region and the greater is the hole concentration at the boundary between the p- and n-regions ($x = 0$).

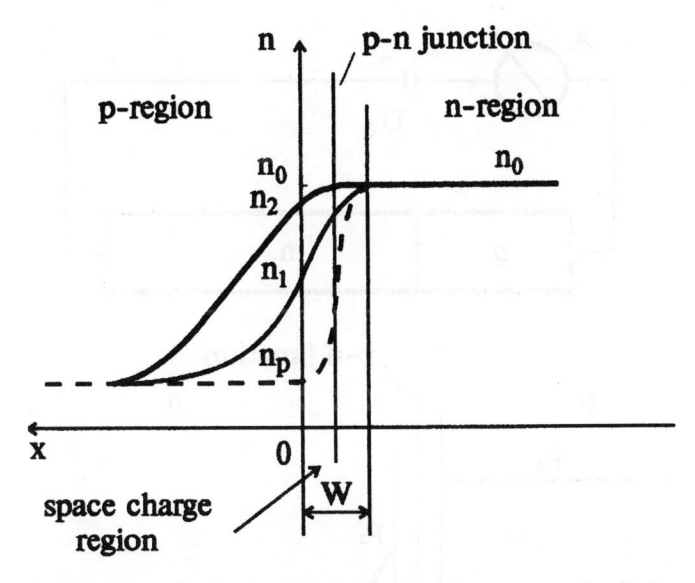

Fig. 62. The distribution of the electrons injected in the p-region of a diode at a forward bias. The greater the forward current, the greater the concentration of electrons injected from n-region into p-region. Compare to Fig. 61.

It is the same with electrons, injected from the n- into the p-region (Fig. 62).

We can see in Figs. 61 and 62 that the greater the forward current, flowing across the diode, the greater the concentration of nonequillibrium minority carriers that are being injected from one region of the junction to another. It is understandable, because the lower the potential barrier, dividing the p- and n-regions, the greater the forward current. And a lower potential barrier corresponds to a greater concentration of the injected carriers.

Figure 63 shows simultaneously the distribution of holes and electrons in the p-n structure when the forward current is flowing across it. As we have already mentioned, in the majority of cases the p-n junctions that are in practical use are abruptly asymmetrical. (One of the regions forming the junction is doped much stronger than the other). In the case indicated in Fig. 63, the p-region is doped stronger.

A very important conclusion can be drawn from Fig. 63. The concentration of carriers that can be injected by a low doped region into a high doped region is always very small, compared to the concentration of carriers in a high doped region. So, in Fig. 63, with the greatest density of the forward current, the

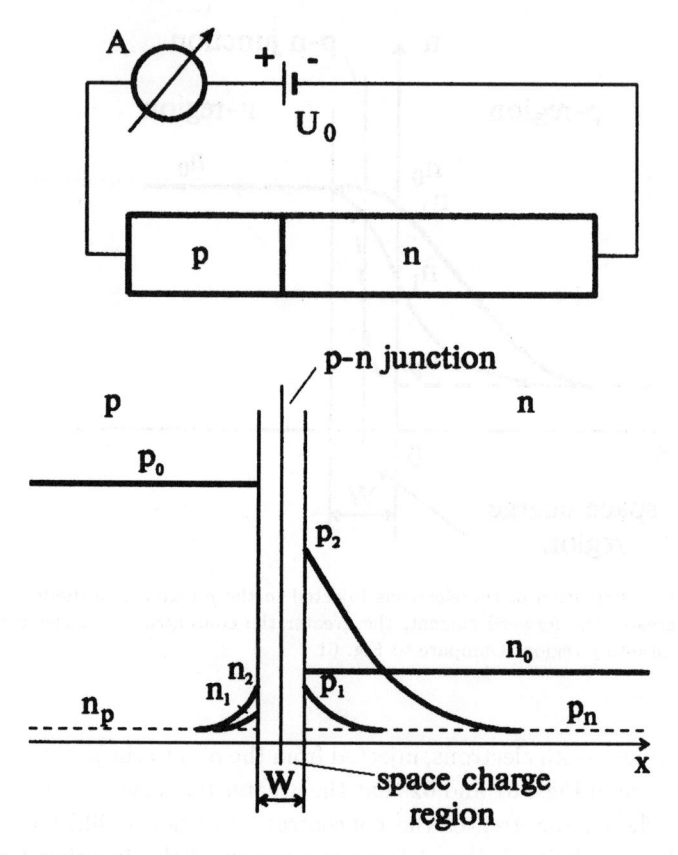

Fig. 63. The distribution of the electrons and holes in a forward biased p-n junction. (For simplicity the $p(x)$ and $n(x)$ dependencies into the space-charge region are not shown). The boundary concentrations p_1 and n_1 correspond to current I_1. The boundary concentrations p_2 and n_2 correspond to current $I_2 (I_2 > I_1)$.

concentration of electrons n injected into the p-region is much less than the value p_0.

But the concentration of carriers which a high doped region can inject into a low doped region even at a relatively small current density, can become many times greater than the initial concentration of carriers in a low doped region. So, in Fig. 63, at the current I_2, the concentration of holes on the boundary of n-region, p_2, is much greater than the concentration of electrons in the n-base $n_0 (p_2 \gg n_0)$.

That is why a high doped region of a diode is called an *emitter* (from the Latin *emmito* - let out). It is capable of injecting such a number of minority carriers into the low doped region, that their concentration will exceed the initial concentration of the majority carriers. The low doped region of a diode is called a *base* (from the Greek — *basis*). As we remember, it is in the low doped region of the junction, the base, that the space charge region is generally located at the reverse bias (Fig. 54). The parameters of the base — its width and level of doping — determine the breakdown voltage of the p-n structure [Eqs. (40), (41)]. So, in accordance with the accepted terminology, Fig. 63 gives the analysis of the operation of a diode with p-emitter and n-base.

Intensity of injection, or as is said, the *injection level* in every region of a diode is usually estimated by the ratio of the concentration of minority carriers on the boundary, where this concentration is maximal, to the concentration of majority carriers. It is seen in Fig. 63 that for the emitter region of the diode the injection level will always be low. The concentration of electrons n for any current is much smaller than p_0.

As for the base region, the situation depends on the current. For the I_1 current, the injection level is low $p_1/n_0 < 1$. And for a greater current I_2, a very high injection level is realized in the base: $p_2/n_0 \gg 1$.

Let us examine very attentively how the current flows across the forward-biased diode.

The length of the n-base of the diode, indicated in Fig. 63 exceeds the diffusion length of holes L_p several times, so that there are no holes in the depth of n-base, far from the p-n junction. All the holes, injected from the p-emitter, die before they reached the right-hand part of the base, which is at the longest distance from the junction. It is clear therefore that the current in the depth of the n-base, far from the junction, can be conducted only by electrons.

If the current I flows across the diode, that means that every second the source of bias U_0 "pushes" into the base $N = I/q$ electrons. These electrons move against the direction of the current, from right to left. And ... what happens to them?

Should the base be just a piece of a semiconductor of n-type, as is shown in Fig. 21, everything would be simple. The same number of electrons that have entered the specimen on the right would have left it on the left during the same time, returning to the source of bias U_0.

In the base of the diode things are quite different. The electron component of the total current I_n flows from the base into the emitter. But as we know now, if the emitter is doped much stronger than the base, then this electron component makes a negligible fraction of the total current I, i.e. a negligibly small part of electrons which have entered the base is leaving it. [See Eqs. (46), (47) and (37), (38)]. From these equations it follows that $I_n/I \approx I_n/I_p \approx j_{ns}/j_{ps} \approx N_d/N_a$. For a very asymmetrical p-n junction the ratio of the doping level of the base to the doping level of the emitter N_d/N_a makes $10^{-4} \div 10^{-5}$.

Thus of 10 000 or even 100 000 electrons which have entered the base only one electron enters the emitter after it has left the base. This tiny fraction is very often neglected and it is said that the electrons from the base do not get into the emitter.

Where do they get to?

They perish, or they recombine with the holes.

All the holes injected from the emitter into the base, if the base is long enough — perish (Fig. 63). But in order to recombine, the hole needs a partner, an electron (Fig. 4). Therefore the number of electrons that must be delivered to the base by the bias source U_0, should be equal to the number of holes dying every second. As we know, in the base near the p-n junction, practically the whole current is transferred by holes, i.e. every second about $N = I/q$ holes from the emitter enter the base. And the same number of electrons comes from the depth of the n-base to meet them.

So, under the action of the electric field, made by the bias source U_0, holes move from the left to the right from the contact to the p-emitter in the direction of the p-n junction and are injected into the n-base. Having got into the n-base, they move, due to diffusion into the depth of the base, recombining on their way with the electrons they come across.

It can be said in other words. Every second, "obeying the push" from the side of the "minus" source, $N = I/q$ electrons enter the base. They are unable to leave the base. Attracted by the negatively charged electrons, the holes enter from the side of the emitter, and perish, recombining with the electrons. To maintain their steady delivery into the base, every second $N = I/q$ electrons come from the p-region into the metal contact, leaving in the p-emitter the same number of holes. That is how interesting and complicate the current flows across the forward-biased diode.

5.3 Summary

Between the parts of the semiconductor crystal, one of which is doped by a donor impurity, the other by the acceptor impurity, there appears a potential barrier — the p-n junction. If there is no external voltage, the height of that barrier approximates the width of the forbidden band of the semiconductor E_g.

The external bias, whose "plus" is applied to the n-region of the junction, and "minus" to the p-region (the reverse voltage), increases the height of the barrier. As this takes place, a very small reverse current flows across the p-n junction. The diode with a p-n junction presents a very large resistance. When there is a forward bias (the "plus" of the external voltage is applied to the p-region of the diode, and the "minus" to the n-region), the height of the barrier is diminished. The forward current density increases abruptly with the increase of the bias and even with a relatively small voltage it can reach very high values. A diode with a p-n junction presents in this case a very small resistance.

Chapter 6

Diodes With The *p-n* Junctions

The properties of the *p-n* junctions which we discussed in the previous chapter underlie the work on dozens of types of the *p-n* diodes. Each of such diode is a semiconductor device which has two external electrodes and a *p-n* junction inside. *P-n* diodes can be used for many different purposes, depending on their construction, on the semiconductor material the *p-n* junction is made of, on the level of doping of the *p* and *n* regions of the junction and on the bias voltage applied. They are used to "rectify" the alternating current, i.e. to transform the alternating current into the direct current (the rectifier diodes). They are used to transform the solar light into the electric current (solar cells). They are also used for the generation, gain and transformation of electric signals (the IMPATT diodes, parametric diodes, tunnel diodes, multiplier of frequency). They are used in order to register and investigate various light signals (photodiodes) and to transform the electric current into light (light-emitted diodes and the semiconductor lasers), and also for many other purposes.

To tell you of all the types of the *p-n* diodes, it would be necessary to write a book much thicker than the one you are reading now.

In this chapter we will discuss just a few types of *p-n* diodes. All of them have an important practical application, and, besides, they vividly illustrate the properties of the *p-n* junction which we studied in Chapter 5.

6.1 Photodiodes

In the myths and fairy tales of various cultures, there are creatures possessing an ability to see in the darkness, able to spot a gnat at a distance of seven miles, or a star hidden behind a thick layer of clouds. Nowadays there are hundreds of devices fulfilling tasks that are not less difficult and important. Special photoelectric security systems are guarding certain objects day and night all the year round. When the beam of light surrounding the guarded territory is interrupted, the security device will react instantly, sounding alarm. Hundreds of times a day the worker takes a sheet of metal, places it under a punch and presses the pedal with his foot. A multi-ton weight presses the slab. The ready part is removed and the new slap is put under the punch. But the worker is a man, he is not a robot, he may get careless and press the pedal without taking his hand away from the danger zone. But nothing happens! The punch does not move. If the hand is in the danger zone, it crosses the beam of light and the photoelectric security device blocks the mechanism which drives the punch.

Photoelectric devices protect buildings from fires, discover the presence of noxious gases and vapours, and analyse the transparency of man's blood. They check the state of bridges and dams, control the chemical and mechanical processes, control and maintain the given level of the liquid, determine the parameters of particles in nuclear physics, count up the parts of devices, sorting them by their shape, colour, mass and size, check the temperature and carry out hundreds of other operations with such a speed, exactness and sensitivity that it leaves far behind the ability of a man's eye. The work of any photoelectric device is based on an element, whose electric parameters — such as resistance, current and voltage — change under illumination. Photodiode is one of the most common devices of this type.

Photodiode is a reverse biased p-n junction. (See Sec. 5.2.4.).

We remember that if the energy of a photon E_{ph} is larger than the forbidden gap E_g, then every photon absorbed generates in the semiconductor an electron-hole pair (See Fig. 4). If the electron and hole appeared under the action of light in the space-charge region of the reverse-biased *p-n* junction, then ... (we can return to *pp. 131–132* and copy the text word for word). "... any carrier, an electron or hole, born in the space charge region, will be immediately caught up by the electric field and be thrown away: the electrons, to the *n*-region; holes, to the *p*-region."

In the darkness the generation current of the *p-n* junction (the so-called dark current) is defined by the number of electrons and holes appearing in the space charge region on account of thermal generation. When the reverse-biased junction is illuminated by light with the quantum energy $E_{ph} > E_g$, the current across the junction increases. The photocurrent which appeared under the action of light is as many times larger than the dark current, as the number of carriers, created by light in the space-charge region, is larger than that of

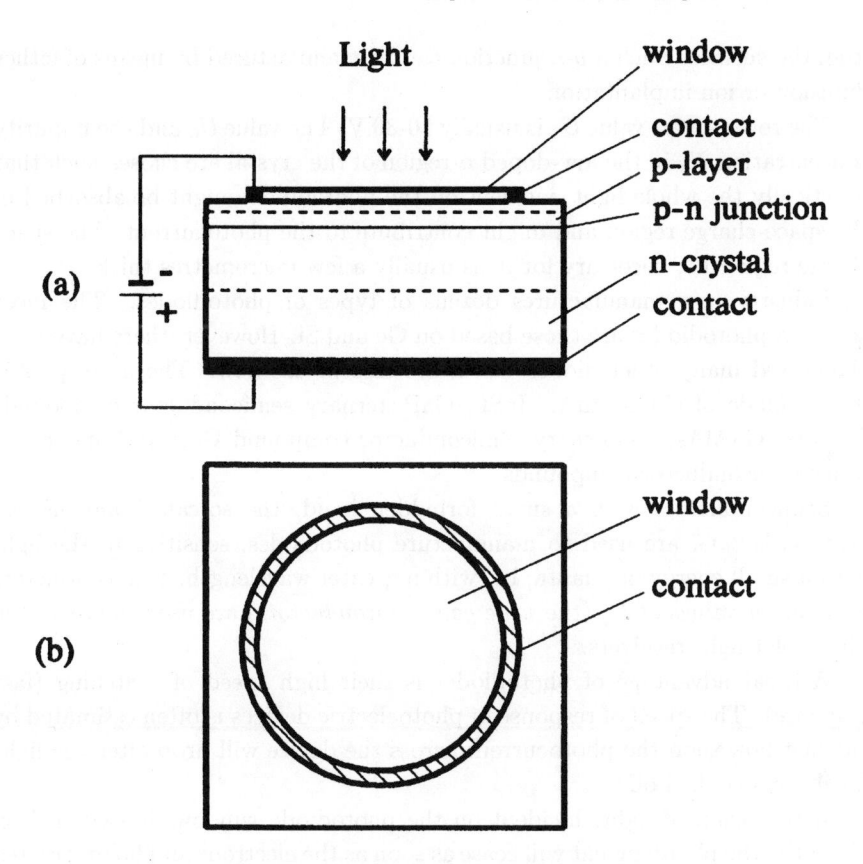

Fig. 64. Design of the typical photodiode.

electrons and holes which appear on account of thermal generation. The photocurrent, amplified and transformed by the electronic circuit, actuates the alarm signals, blocking devices and other guard systems.

Figure 64(a) shows a typical design of a photodiode. We can see that it is an old acquaintance of ours: it is a *p-n* junction, biased in the reverse direction. The upper contact to diode is made in the shape of a ring [Fig. 64(b)]. The region of the *p-n* junction inside the ring, the so-called window, is accessible to light. In the photodiode the plane of the *p-n* junction is located close to the surface of the crystal; lest the light, incident on the crystal should be absorbed on the way to the space charge region. Usually, the plane of the *p-n* junction in photodiodes is located at a distance of several decimal fractions of micrometer from the surface. Such a p-n junction can be manufactured by means of either diffusion or ion implantation.

The reverse bias value U_0 is usually 10-30 V. The value U_0 and the impurity concentration N_d in the low-doped n-region of the crystal are chosen such that practically the whole light, incident on the photodiode, might be absorbed in the space-charge region and might contribute to the photocurrent. The space charge region W, necessary for it, is usually a few micrometres thick.

Industry now manufactures dozens of types of photodiodes. The most common photodiodes are those based on Ge and Si. However, there have been elaborated many other diodes which are also in use now. Those are photodiodes made of GaAs, InAs, InSb, GaP, ternary semiconductor compounds GaInAs, GaAlAs, quaternary semiconductor compound GaInAsP and many other semiconductor compounds.

Semiconductors with a small forbidden band, the so-called *narrow gap semiconductors*, are used to manufacture photodiodes, sensitive to the light with a small energy of guanta, i.e. with a greater wavelength. Semiconductors with larger values of E_g (the *wide gap semiconductors*) are used to create the ultraviolet light receivers.

A great advantage of photodiodes is their high speed of switching (fast response). The speed of response of photoelectric devices is often estimated by the fact how soon the photocurrent across the device will drop after the light has been switched off.

If the beam of light, incident on the photodiode window, is switched off abruptly, the photocurrent will cease as soon as the electrons and holes, created by light, are carried by the field away from the space charge region.

The typical values of the electric field in the reverse biased p-n junctions are large. They are $\sim 10^4$–10^5 V/cm. In such strong fields the drift velocity of electrons and holes v does not depend on the field and is $v_s \sim 10^7$ cm/s (See Fig. 20).

The way, the electrons and holes are covered before they are carried by the field away from the p-n junction, and this is evidently equal to the width of the space charge region W. As was already mentioned, the value W for typical photodiodes is several micrometres.

Thus, the time, determining the speed of response of the photodiode, is approximately

$$t_0 \sim \frac{W}{v_s} \sim 10^{-10} \div 10^{-11} s$$

The photodiode is capable of reacting to the light intensity change as fast as hundred milliard fractions of a second!

The fast response makes it possible to use photodiodes in the highest speed systems of transmitting and processing information, for instance in systems of fibre optics communication. It is photodiodes that serve as receivers of the fast modulated signals in such systems.

Nowadays, scientists work at the creation of multi-element photoreceivers, possessing the "colour" eyesight. The design of such devices resembles the structure of the human retina. Each of dozens or hundreds of thousands of elements, form the artificial eye, emitting electric signals, corresponding to the intensity and wavelength of the light incident on it. Such devices make it possible to put the most complicated data into computers without any special coding; they can directly let the computer know what they actually "see".

6.2 Variable Capacitors

Variable capacitors use the property of the reverse biased p-n junction to change the capacity when the bias is changed. The greater the bias, the smaller the capacity [See Eq. (44)]. Practically everyone has to deal with capacitors. Everyone who tunes his radio set to the station he wants to get, changes the capacitance of the variable capacitor.

Without variable capacitors no TV transmission would be possible, especially that of colour TV.

The semiconductor capacitances, controlled by bias voltage, underlie the basis of supersensitive transformers changing the direct voltage into alternating voltage. Due to them the amplification of superweak signals is realised, including the signals from cosmic space (Fig. 65). They are used to generate signals with a frequency of hundreds of milliards of Hertz.

All these applications of the variable semiconductor capacitance are based on the fact that the value of this capacitance can be changed very rapidly by means of bias.

The capacitance of the variable capacitor changes, depending on its construction, in thousandth, millionth and sometimes milliardth fractions of a second, when the bias is changed.

One of the most important parameters of a variable capacitor is the capacitance change coefficient K_c, defined as the ratio of the maximal capacitance

Fig. 65. Parabolic reflector antenna of power radiotelescope. Signals received from the remote regions of the Universe even by such enormous antennas are inconceivably week. To amplify them supersensitive amplifiers are required. The main element of such amplifiers is a variable capacitor (varicap).

of the variable capacitor $C_{\max}$ to its minimal capacitance $C_{\min}$: $K_c = C_{\max}/C_{\min}$.

Since the capacitance of the variable capacitor depends on bias, it is clear that the value K_c depends on the possible range of the reverse bias applied to the variable capacitor.

The main condition, limiting this range, is formulated quite simply: within the operational bias range of the variable capacitor the current across it must not be large.

The minimal reverse bias U_0 at the variable capacitor should as a rule be equal to zero. Then the voltage drop at the p-n junction is minimal and is equal to U_{pn}, while the capacitance of the variable capacitor is maximal.

According to Eqs. (44) and (40), it is as follows

$$C_{max} = S \left(\frac{q \, \varepsilon \, \varepsilon_0 \, N_d}{2 \, U_{pn}} \right)^{1/2} \qquad (49)$$

The maximal bias U_i is limited by the process of impact ionization. Assuming that $U_i \gg U_{pn}$, we obtain from Eq. (41) the value of the minimal capacitance of the variable capacitor.

$$C_{min} = S \left(\frac{q \, \varepsilon \, \varepsilon_0 \, N_d}{2 \, U_i} \right)^{1/2} = S \frac{q \, N_d}{F_i} \qquad (50)$$

Thus, the capacitance change coefficient Kc will be equal to

$$K_c = \frac{C_{max}}{C_{min}} = \left(\frac{\varepsilon \, \varepsilon_0 \, F_i^2}{2 \, q \, U_{pn} \, N_d} \right)^{1/2} \qquad (51)$$

It is seen that the smaller the impurity concentration in the low-doped region, the larger K_c becomes. In practice, the value N_d is chosen to be $10^{15} - 10^{18}$ cm^{-3}. The value K_c in real varicaps is in the range from ~ 2 to 15.

6.3 Light-Emitted Diodes

The basis for the operation of photodiodes and variable capacitors, discussed in the previous sections, were the properties of the reverse biased p-n junction. A very important and demonstrative application of the forward biased p-n junctions is, undoubtably, the light-emitted diodes (LEDs).

It can be said that any forward-biased p-n junction is, in principle, a light-emitted diode. Indeed, the carriers, which from the emitter get into the base, recombine. At least a part of them does, giving birth to a photon (See Fig. 4). A certain part of the photons, which were not absorbed with the diode, will flow outside. That is the light, created from the forward current passing across the p-n junction.

To create a practically suitable light-emitted diode it is necessary to have a certain type of a semiconductor. It would be wrong to think that any semiconductor would do. The first quite obvious requirement is that the greater part of carriers should perish inside it, releasing photons. Or, speaking in a more learned manner, only those materials which possess a rather high probability of *radiation recombination* are suitable for manufacturing LEDs. Germanium and silicon, for instance, do not answer that requirement. The overwhelming majority of electrons and holes in them perish without emitting photons, nonradiative, just recombining via deep centres. While in gallium arsenide (GaAs) and in some ternary semiconductor compounds: GaAsP and GaAlAs, even not very well purified, the probability of radiation recombination is close to a unit. Almost every electron-hole pair perishes, radiating a photon. Therefore a greater part of light-emitted diodes is manufactured nowadays on the basis of these three materials.

The colour, i.e. the wavelength of light, radiated by the light-emitted diode, is defined by the energy of a photon E_{ph} emitted during the recombination. In most cases this energy is close to the energy gap of the semiconductor E_g. So, the wavelength of the gallium arsenide ($E_g \sim 1.4$ eV) light-emitted diode radiation is equal to $\lambda = hc/E_g = 1.24/1.4 = 0.89$ mkm. Such radiation is invisible for the human eye. It corresponds to the infrared region of the spectrum, but that does not hinder at all a very wide use of the gallium arsenide sources of light. Silicon photodiodes, for which the energy of a photon $E_g \sim 1.4$ eV roughly corresponds to the maximum of their sensitivity, serve as receivers of that infrared radiation.

Adding to the lattice GaAs atoms of phosphorus (P) or of aluminium (Al) leads to the increase of energy gap E_g (Fig. 66). With the increase of the content of phosphorus x in the ternary semiconductor compound $GaAs_{1-x}P_x$, the energy gap of the semiconductor increases, as shown in Fig. 66, up to the value $E_g = 2.26$ eV, corresponding to GaP ($x = 1$). Preparing a compound GaAsP with different fractions of P, we obtain the light-emitted diodes, radiating red light. But one fails to prepare in this way yellow or green light-emitted diodes, although the energy of quanta, corresponding to these colours is less than the value of the forbidden band of pure GaP ($E_g = 2.26$ eV). The thing is that when $x \geq 0.45$, the probability of the radiative recombination in GaAsP falls abruptly and the light-emitted diodes are but of low effeciency. In order to increase the efficiency of the radiation, special impurities are added either to the GaAsP or into the pure GaP. These impurities (nitrogen N, zinc Zn, oxygen O) increase the probability of the radiation recombination. Today a wide set of semiconductor materials is used to create "multicolor" LEDs. The use of GaAlAs serves to obtain infrared and red sources of light. LEDs with the colour of light from red to green are manufactured on the basis of the compounds GaAsP and GaP. Quaternary semiconductor compounds GaInAsP makes it possible to create infrared light-emitted diodes whose wavelength is $\lambda \sim 1 \div 1.3$ mkm, which are the best sources for fiber optic lines of communication. And on the basis of the silicon carbide, it is possible to obtain all so-called *main colours*: red, green and blue.

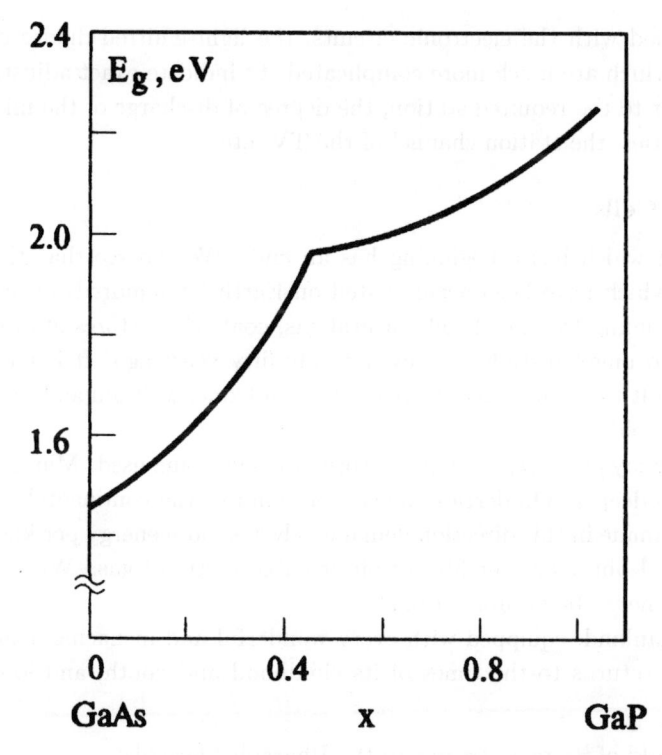

Fig. 66. The band gap dependence of $GaAs_{1-x}P_x$ ternary compound on the fraction of P.

These are called the main colours because mixing them in a certain proportion it is possible to obtain any colour, including white. It is this property of the main colours that determines the work of the colour TV. The screen of a colour TV is covered by three luminophors, radiating red, blue and green light. To cause the radiation of the luminophors, they are excited by electric beams of high energy.

In many laboratories of the world the producers of light-emitted diodes are working at manufacturing a flat colour TV screen. The light-emitted diodes of the main colours may serve the basis of such a screen.

LEDs are chiefly used as various indicators. Nowadays, when there is such an excess of information, the appearance of LEDs — convenient, universal, cheap and effective devices for the indication of information — makes it possible to solve many important problems.

LEDs inform us whether the TV or radio set is on or off, whether the door of the car is shut or open, etc.

Combined with the electronic circuits, the light-emitted diodes can solve problems which are much more complicated: to indicate exact adjustments of the receiver to the required station, the degree of discharge of the microcalculator batteries, the station channel of the TV, etc.

6.4 Solar Cells

Everything which has a beginning has an end. We can see that the energy resources which have been accumulated on Earth for hundreds of millions of years are coming to an end: oil, natural gas, coal. Even stores of uranium do not seem so unexhaustible as they did only fifty years ago. It is increasingly more difficult to get iron and copper, silver and zinc, wolfram and manganese, nickel and gold.

In everyday practice, much poorer ores are now being used. Mines are being dug always deeper. Oil derricks move from land to the continental shelf. But every step made in this direction demands always more energy per kilogramme of the metal obtained, per litre of oil or cubic metre of gas. Where can the necessary energy be obtained from?

And Mankind, equipped with every wonderful and menacing achievement of science, returns to the times of its childhood and youth, and looks up at the Sun.

"The gold of its rays streams to the Pharaohs' nostrils:
May I be bathed in your rays every day?"[a]

Each second the Earth gets from the Sun the energy $\sim 3 \times 10^{17}$ J in the form of radiation. It is 30000 times more than all the power stations of the world — thermal-, hydro- or atomic can produce in the same time. And this wonderful source is inexhaustible — the thermo-nuclear reaction will go on steadily on the Sun for many milliard years.

It would be unforgivable wastefulness not to use such a generous gift of energy. Therefore the ways of transforming the solar energy into the electric energy have been investigated for a very long time. The first proposal to use semiconductors for this aim was made in 1884 when they tried to produce a solar cell based on selenium-gold compounds. More than one hundred years have passed and the work on creating and improving the semiconductor transformers of the solar energy is still going on.

---

[a]This extract from a Egyptian hymn is taken by us from a wonderful book by S.I.Vavilov "The Eye and the Sun". Moscow, Nauka, 1981.

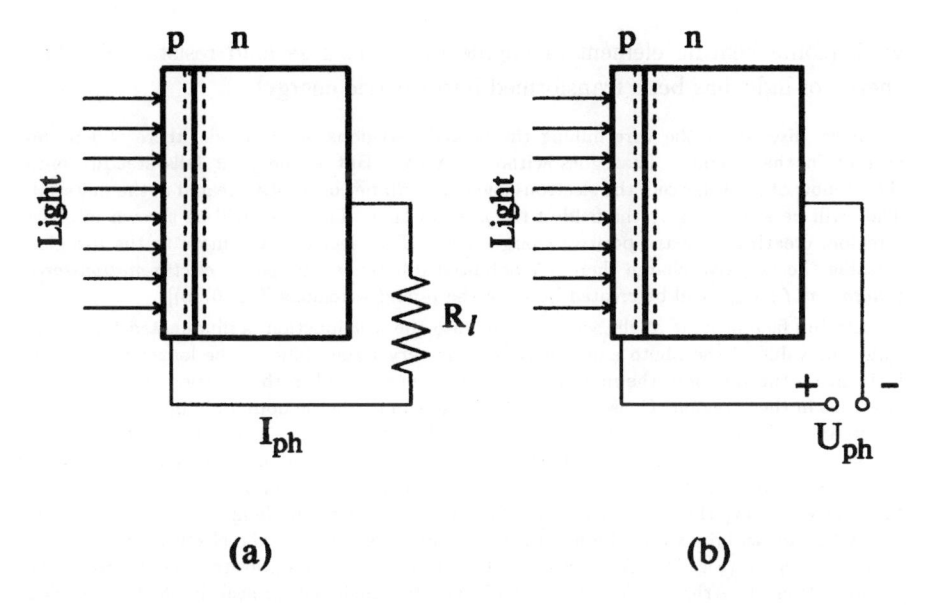

Fig. 67. The generation of photocurrent (a) and photo e.m.f. (b) under illumination.

How are the solar elements constructed and what is the problem of their manufacture?

The solar cell is the conventional *p-n* junction and its work is very much like that of the photodiode which is already well-known to us. But unlike the photodiode, the external bias is not applied to the solar cell (Fig. 67).

Let us come back to Fig. 64 and assume that no external bias has been applied to the diode and there is no bias source in the external circuit. And the bottom (solid) and top (ring-shaped) contacts are connected by the load resistance R_l [Fig. 67(a)].

As we know, even if there is no bias, there exists a potential barrier between the *p*- and *n*-regions. There is a space charge region near the *p-n* junction [Fig. 50(a)] within which there is an electric field [Fig. 50(b)].

That means that even without any external bias the electron-hole pairs which have appeared in the space charge region under illumination, will be separated by the field: the holes will be thrown into the *p*-region, and the electrons, into the *n*-region. Their displacement will create an electric current in the external circuit, the photocurrent, directed from the *p*- to the *n*-region. The photocurrent I_{ph}, flowing across the load resistance R_l, performs useful work (some heating element or engine being used as load resistance). The energy of light has been transformed into electric energy!

If we disconnect the wire linking the p- and n-regions of the diode, there will be no current in the circuit. That goes without saying. But if the light falls steadily onto the window of the solar cell, the electron-hole pairs will be constantly created in the material. They will be separated by the field of the p-n junction. The holes will be thrown into the p-region, creating an extra positive charge there. The electrons will move to the n-region creating the negative charge there. A potential difference, the *photo electro-motion-force (photo e.m.f.)* U_{ph}, will be created between the p- and n-regions [Fig. 67(b)].

On the face of it, it might seem that in case the p-n junction is illuminated for a long time, the value of the photo e.m.f. may become very large. Indeed, the longer the light is incident on the junction, the more holes will be accumulated in the p-region, and the more electrons in the n-region. Consequently, the greater must be the potential difference between the p- and n-layers of the junction ... But in reality it is not so. The polarity of the photo e.m.f. - the "plus" on the p-side and the "minus" on the n-side of the region corresponds, as can be easily seen, to the forward bias on the p-n junction [compare Figs. 67(b), 58 and 63]. As we know, the forward bias on the junction lowers the height of the barrier and facilitates the withdrawal of the holes from p- into n-region, and the electrons from the n- into the p-region (Fig. 58). As a result of the action of the two opposite mechanisms — the accumulation of carriers under the effect of light and their withdrawal due to the lowering of the height of the barrier at every value of the intensity of light incident on the cell, a certain value of the photo e.m.f. U_{ph} is established in the steady state. With the increase of the intensity of light the value U_{ph} grows very weakly, proportionally to the logarithm of intensity. With the intensity of light being the greatest, when under the action of the photo e.m.f. the barrier turns to be practically "rectified", the value $U_{ph\ max} \approx U_{pn} = E_g/q$.

As we see, the principle of the action of the solar cells is very simple. Then why are so many people all over the world toiling trying to elaborate various solar cells and batteries? They do their best to improve the reliability and durability of the cells, to reduce their cost and, mainly, to improve their efficiency. If every quantum of sunlight, incident on the solar cell, could be transformed into an electron-hole pair, contributing to the photocurrent, then this hypothetic solar cell would have a 100% efficiency.

The efficiency of the first solar cells was some hundredth fractions of percent. The efficiency of modern cells has grown thousands of times and makes 10-20%.

The main difficulty on the way to increasing the solar cell efficiency lies in the fact that the sunlight consists of photons of various energies. It comprises quanta corresponding to the ultra-violet light ($E_{ph} \geq 3$ eV), and quanta of the visible light and infrared radiation ($E_{ph} \leq 1.5$ eV). Meanwhile for every solar cell, no matter what kind of semiconductor it is made of, there is spectral sensitivity curve (Fig. 68).

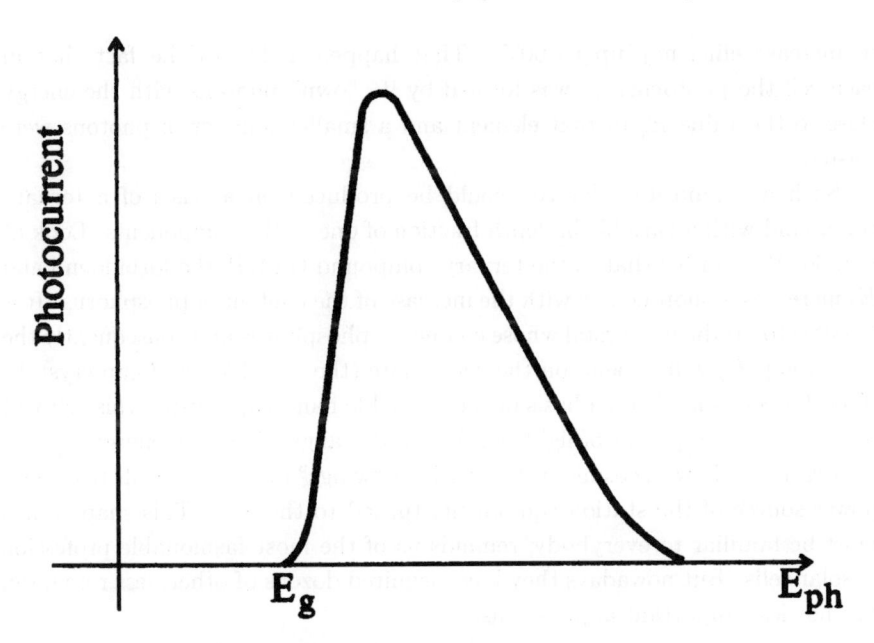

Fig. 68. The qualitative dependence of photocurrent on the photon energy.

This curve shows that only the quanta of light with the energy $E_{ph} \sim E_g$ can be efficiently transformed into the photocurrent.

If the material for producing the cell chosen has a wide forbidden band E_g, then practically the whole spectrum of sunlight will pass through that cell without creating any electron-hole pairs ($E_{ph} < E_g$).

If, on the contrary, the material for producing the cell will have a small value of E_g, then practically all the photons in the sunlight will have the energy $E_{ph} \gg E_g$. Figure 68 shows quite clearly that such photons contribute very little to the photocurrent. They are absorbed at the very surface of the cell, far from the p-n junction, and the electron-hole pairs, created by them, perish very soon, at the expense of the surface recombination.

Calculations have shown that the cells made of the semiconductor with $E_g \approx 1.5$ eV must have the maximum efficiency of transforming the energy of sunlight into electric energy. The values of the forbidden band of the well-studied materials — Si and GaAs — are quite close to that optimal value E_g. Solar cells are mainly made from them.

Many new methods have been proposed to increase the maximum efficiency. One of the ways was to connect in series cells with different E_g which helped to increase efficiency up to 60%. That happened due to the fact that in each cell the photocurrent was formed by its "own" photons with the energy close to the value E_g of that element and a smaller number of photons were wasted.

Such a composite solar cell could be produced on a basis of a ternary compound with a variable in depth fraction of one of the components. Look at Fig. 66. We can see that in the ternary compound GaAsP, the forbidden band E_g increases monotonously with the increase of the content of phosphorus. It is possible to produce a crystal whose content of phosphorus and consequently the energy gap E_g will depend on the coordinate (the variable band gap crystal). If a solar cell is made on a basis of such variable band gap crystal, this cell will be able to catch quanta of light within a wide range of photon energy E_{ph}.

Figure 69 shows a cosmic station with its "wings" of the solar cells (the main power source of the station equipment) turned to the Sun. This snap, which must be familiar to everybody, reminds us of the most fashionable profession of solar cells. But nowadays they have acquired dozens of other, less romantic, but not less important applications.

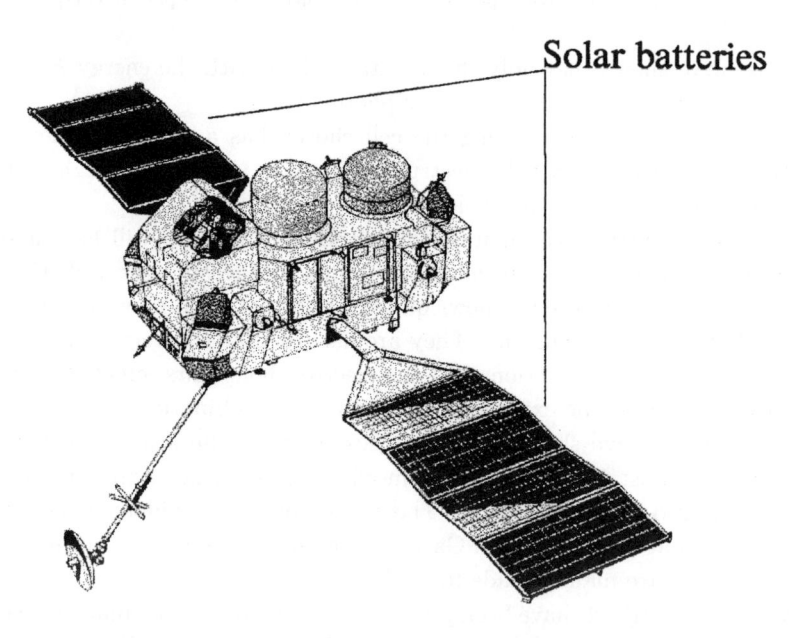

Fig. 69. Solar batteries serve as an energy source for satellites.

In deserts, the solar cells are used to pump out water from deep aquifers. The solar cells illuminate and heat the houses, they rotate the engines, charge accumulator batteries, etc.

6.5 Rectifier Diodes

Practically all the electric energy consumed in the world is generated nowadays by turbo-generators of power stations in the form of an alternating voltage at the so-called industrial frequency (50 or 60 Hz). But in many cases, it is quite impossible to use the energy in that form. Many devices, machines and even whole branches of industry are in need not of the alternating, but of the direct current.

The direct current feeds the underground and overland electric transport. To provide the work of this transport it is necessary to transform the alternating voltage into direct voltage — to rectify currents of hundreds of amperes at the voltage of hundreds and thousands of volts.

Powerful electrolysis stations which consume currents of hundreds of thousands and even of millions of amperes at a rather small voltage must have direct current. Powerful modern supercomputers also want large direct current at the voltage of several volts. On the other hand, in radiotechnique and in microwave electronics, it is often necessary to rectify signals whose voltage is millionth or even billionth fractions of a volt and whose frequency is of fractions of Hz (in seismology) to thousands of billions of Hz in microwave electronics. The rectifying diodes with the p-n junction cope quite successfully with all those complicated problems.

The principle of transforming the alternating current into the direct current is illustrated in Fig. 70(a). The switch K is connected in series with the load resistance R_l (R_l may be any object, operated at the direct current, such as accumulator, computer, electrolysis station, etc.). The alternating voltage $U(t)$ is delivered to the input of the circuit between the points A and B. When the voltage at the input is positive, the switch will be kept closed. When the voltage at the input changes its sign, the switch will be kept open [Fig. 70(b)]. As a result of it, the dependence of the current across the load resistance R_l in time will look the way it is shown in Fig. 70(c). When the switch is closed, the current $I = U/R_l$ is flowing across the circuit. When the switch is open, the current in the circuit is equal to zero and all the voltage is applied to the switch — the place where the circuit is broken. As it is said, the switch blocks the voltage applied.

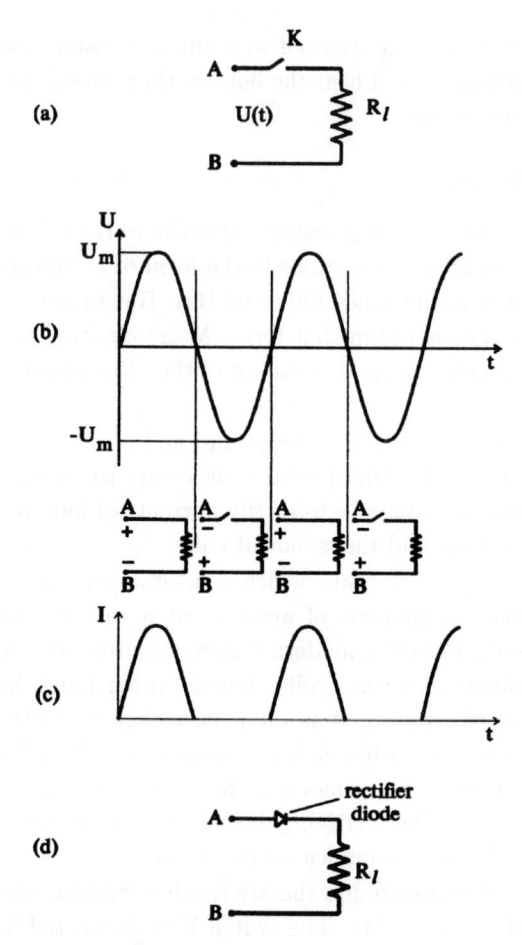

160 Transistors. From Crystals to Integrated Circuits

Fig. 70. Circuit diagrams of rectifing the alternating current. (a) The switch K is connected in series with the load resistance R_l. (b) When the voltage at the input is positive, the switch will be kept closed. When the voltage at the input changes its sign, the switch will be kept open. (c) A current I accross R_l always flows in the same direction: from A to B. (d) The rectifier diode with a p-n junction acts as a switch K in real circuits.

The current whose dependence on time is shown in Fig. 70(c) is no longer alternating. It always flows in the same direction — from A to B. Though it can hardly be called direct either: its amplitude changes greatly, it pulses and the current is called the pulse current. This pulse current is quite fit for certain purposes: to charge accumulator batteries, or for electrolysis. If the too large pulsation is unwanted, it can be smoothed by means of one of the numerous circuits of the so-called filters. The simplest filter is a capacitor, connected in parallel with the resistance R_l.

Now, if instead of the hypothetical switch K we use a diode with a *p-n* junction [Fig. 70(d)], we will get a true rectifier circuit, commonly used in practice.[b] When the voltage at the diode corresponds to forward bias, its resistance is very small. The current in the circuit is $I \approx U/R_l$. When the voltage changes sign and becomes negative, the diode proves to be reverse biased, its resistance being very large. The diode blocks the voltage applied, and the current in the circuit is very small.

Let us see now, what requirements should the diode with a *p-n* junction satisfy, so as to operate successfully as a rectifier.

It is clear that first of all that diode should conduct in the forward direction such a current which is wanted for the load resistance to be in operation. If that current is equal to several milliamperes, as it is when feeding a transistor receiver or a microcalculator, it is not difficult to have that requirement satisfied. But as we have seen, the current which is wanted, may be hundreds, thousands or even hundreds of thousands of amperes.

The stronger the current, which rectifies the diode, the greater the area it must have. This requirement stems from a simple fact that the diode is heated in the process of operation. The energy, released in the diode as heat, is, naturally, the greater, the greater is the current, flowing across the diode. And the energy which can be released by the diode into the surrounding space is proportional to its area.

For the silicon *p-n* junctions (the overwhelming majority of rectifying diodes are made of silicon) the allowable density of the forward current, calculated from the conditions of heat removing is $\approx 50 \div 100$ A/cm^2.

The area of the power rectifier diodes makes tens of cm^2. Such diodes are able to rectify currents of thousands of amperes. If it is necessary to rectify still greater currents, the diodes are connected in parallel.

The second requirement, which the rectifier diode must meet, is its ability to endure, without breaking down, the maximum value of the voltage U_m which is to be rectified. As we know, in order to meet this requirement, first, the impurity concentration in the low doped region of the diode — the base — must be small enough [see Eq. (41)]. And secondly, the width of the base W must be large enough, so that when the reverse voltage at the diode is

---

[b]Pay attention to the conventional designation of the *p-n* diode in the circuit. The *n*-region always corresponds to the vertex of the triangle, designating the diode, while the *p*-region corresponds to its base. The diode is forward biased, and it conducts the current, when "plus" is applied to the base of triangle. The diode is biased in the reverse direction and current is equal to zero when "minus" is applied to the base of the triangle.

maximal, it must be greater than (or, at least equal to) the width of the space charge region W_1 [Eq. (40), Fig. 54(b),(c)].

So, in order to operate as an open-circuit switch and to endure a large reverse voltage, the diode must have a long low doped base. But the long base, made of a semiconductor with a low concentration of electrons, will have a large resistance! With this large resistance in that part of the period when the diode must work as a short switch, conducting a strong current, a considerable voltage will drop. And that cannot be tolerated.

First of all, the voltage, which drops at the switch decreases the useful voltage at the load resistance R_l. And besides, when the voltage drop and the current at the switch are large, too much heat is released.

Let us make sure that for high-voltage devices the discrepancy we paid attention to is rather serious. Discussing Eq. (41), we have already established that for the diode to block the voltage $U_m \sim 3000$ V, the doping impurity concentration in the base should not exceed $N \sim 10^{14}$ cm^{-3}. Substituting the values $U_0 = 3 \cdot 10^3$ V and $N = 10^{20}$ m^{-3} into Eq. (40), we obtain that the width of the base must exceed the value $W_1 \sim 2 \cdot 10^{-4}$ m=2×10^{-2} cm. As was mentioned, the forward current density in the rectifier diodes is $j = 50 \div 100$ A/cm^2.

And now we make a simple calculation based on Ohm's law. As we know, $j = qnv = qn\mu F$. Let the rectifier diode be made of silicon, with the diode base of n-type. The mobility of electrons is $\mu_n = 0.13$ m^2/(V · s) (See Table 3). The electron concentration n is equal to the doping impurity concentration N. The field in the diode base $F = j/qN\mu = (2.5 \div 5) \cdot 10^3$ V/cm. The voltage drop at the diode is $U_B = FW = 50 - 100$ V!

That's a nice "switch", at which, when short, 100 volts drop! It goes without saying that in practice such a switch does not serve any purpose.

The situation seems hopeless. If you want to have the high alternating voltage rectified, you have to put up with the tremendous voltage drop at the diode in forward direction. But nevertheless there is a way out. In fact, we even know it. It consists in using the phenomenon of injection.

Let us look once more at Fig. 63. We can see that when the injection level is high, the carrier concentration in the base close to the p-n junction can be much higher than the equillibrium concentration of carriers n_0. The higher the concentration of carriers, the smaller is the base resistance and, consequently, the smaller voltage will drop on the base, the current density being the same. Is it possible to construct a diode in such a way that with the forward current flowing across it, the whole base might be filled with nonequillibrium carriers, injected from the emitter, with the resistance of the base decreasing abruptly?

It is. It is quite possible. And it is easy to see what is required for it. It is necessary that the base of the diode should not be too long, lest the concentration of holes injected by the emitter should decrease too abruptly at the expense of recombination to the opposite end of the base.

But the width of the base W cannot be chosen arbitrarily! As we have seen, it is determined by the maximal voltage U_m which must be blocked by the diode. For large values of U_m the width of the base must also be large.

But ... please pay attention to the fact that how much the concentration of non-equilibrium carriers falls into the depth of the base depends not on the value W as such but on the relation W/L_p. It depends on the numbes of diffusion lengths of holes L_p that can be placed along the base W. Even if the value W is large, and the relation W/L_p is small, the fall of the concentration will be not quite appreciable. Therefore, the larger the voltage being blocked, the longer the base, the greater must be the diffusion length of holes L_p, if we want to make use of the phenomenon of injection and obtain an acceptable forward voltage drop with the forward current densities being large.

Assuming that the diffusion length of holes must be of the width of the base ($W \approx L_p$), and making use of Eq. (26), it is easy to establish that for the silicon diode whose width of the base is $W = 200$ μm the lifetime of carriers in the base must be

$$\tau_p \gtrsim \frac{L_p^2}{D_p} \approx \frac{W^2}{D_p} \gtrsim 4.10^{-5} s\,.$$

(The value of D_p for silicon ≈ 10 cm^2/s.).

As we know, the increase of the lifetime of carriers is achieved by removing the deep impurities from the semiconductor (Chap. 2). Technologists can remove impurities from silicon very well and the material whose lifetime $\tau_p \geq 40$ μs is quite available.

Moreover, doing one's best, one can obtain silicon whose lifetime is several hundred microseconds or even several milliseconds. Does it mean that we must try to make the rectifier diodes for the voltage of tens of thousands of volts, using such silicon as a base?

Let us not hurry to give the answer. Any switch, used to rectify the alternating current, is characterised by one more parameter, about which we preferred not to consider for a while. It is a speed of response. The diode must turn from the state "on" to the state "off" in a time far shorter than the period of oscillations of the current. The period $T = 1/f \approx 15$ ms corresponds to the industrial frequency of 60 Hz. If we assume that the switch should be turned from the "on" to "off" state during the time of the order of 1/100 of the oscillation period T, then the rectifier diode should be able to pass from the state "on" to "off" and vice versa during the time not more than $\approx$ 150 µs. But what is it that determines the time of switching the semiconductor diode? That is the lifetime of the minor carriers within the base! Indeed, the transition from the state "off" to the state "on" corresponds to the transition from the state when there are no minor carriers in the base to the state when the carriers injected from the emitter fill up all the base. When the reverse voltage at the diode is changed for the forward voltage, the carriers (holes) begin moving from the emitter to the base at the expense of diffusion and fill up the base during the time $t \sim W^2/D_p$ [See Eq. (25)]. But since $W^2 \sim L_p^2$, then

$$t \approx \frac{L_p^2}{D_p} \approx \tau_p$$

The time of the diode being switched "on" proves to be of the order of lifetime of the minor carriers in the base.

When the opposite takes place and the voltage at the diode is changed from forward to the reverse, for the diode to pass to the state "off" it is necessary that the holes, which at the forward voltage filled up the base, should dissappear (recombine). Recombination time of the carriers is their lifetime τ_p. Thus, the time of switching from the "on" to "off" state proves to be equal to the lifetime of minor carriers.

It is clear now why the greater the voltage the rectifier diode is designed for, the smaller the limiting frequency of oscillations it is able to rectify efficiently. The greater the voltage, the longer the base. The longer the base, the greater must be the lifetime of carriers τ. The greater τ is, the slower the switching ("on" and "off") of the diode becomes. The limiting voltage of the diodes designed for the work in the circuits of the industrial frequency, is roughly 10^4 V.

But what is to be done if it is necessary to rectify a greater voltage, of, say, hundreds of thousands or millions of volts? In that case, the rectifier diodes are to be connected in series.

6.6 Summary

We have considered several examples of how the properties of the p-n junction are used to create various diodes. We have seen how by means of the p-n junction light can be transformed into electric current, and the electric current, into light. We have seen how it is possible due to a p-n junction to change the frequency and to rectify the alternating current, transforming it into the direct current. But in fact the sphere of the application of the p-n diodes is much wider. Moreover, every year new semiconductor devices are being proposed, which are based on the properties of the p-n junctions. All these applications are based on the same physical principles of the operation of a p-n junction, which we discussed in Chapter 5.

Part III

Transistors

Transistors were invented in 1948 by American physicists J. Bardeen, W. Brattain, and W. Shockley. In 1956, they were jointly awarded the Nobel Prize.

This creation by the American scientists was called *Transistor*. But in 1952, one of the Nobel Prize winners elaborated on another type of semiconductor transistor whose principle of operation differed a lot from the first transistor. That transistor, invented by W. Shockley was named the *field-effect transistor* (*FET*), while the one invented by Bardeen, Brattain and Shockley was called the *bipolar transistor* (*BT*).

Though the physical principles of operation of those two transistors are quite different, both of them are meant for the same purpose — for amplifying the electric signals, and that has almost innumerable practical applications. The signal which arrives at the input of the transistor gives rise to a much more powerful signal at the output.

In Chapter 7, we will get to know the principles of work of the bipolar transistor. Chapter 8 contains the physical principles of operation of the field-effect transistor, and Chapter 9 gives the description of some applications of these wonderful devices.

Chapter 7

Bipolar Transistors

The arrangement of a bipolar transistor is shown schematically in Fig. 71. It is quite simple. Between the two p-type regions there is a narrow region of the n-type semiconductor. That is why it is called the p-n-p transistor. The names of the two parts of the structure are known to us from the previous chapter: *emitter* and *base*. The name of the third part is the *collector*. In a similar way one can obtain a transistor of the n-p-n type. Then a narrow strip of a p-type semiconductor is to be placed between the two n-regions. The physical principles of operation of both structures (p-n-p and n-p-n) are absolutely the same. If one gets a good idea of the work of a p-n-p transistor, he or she will be able to quite easily analyse the work of the n-p-n structure.

The structure shown in Fig. 71 can be described in another way, and namely as a p-n junction which is quite familiar to us, to which one more p-region has been added. Or, otherwise, as two p-n junctions located very close to each other, and having a common base. How can this simple structure fulfill its main mission — to amplify the electric signals?

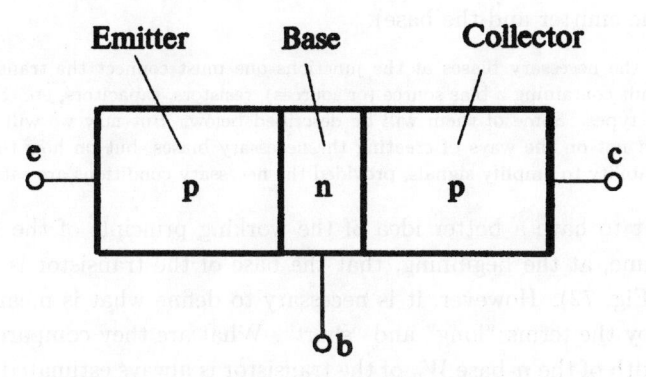

Fig. 71. Schematic diagram of the bipolar p-n-p transistor.

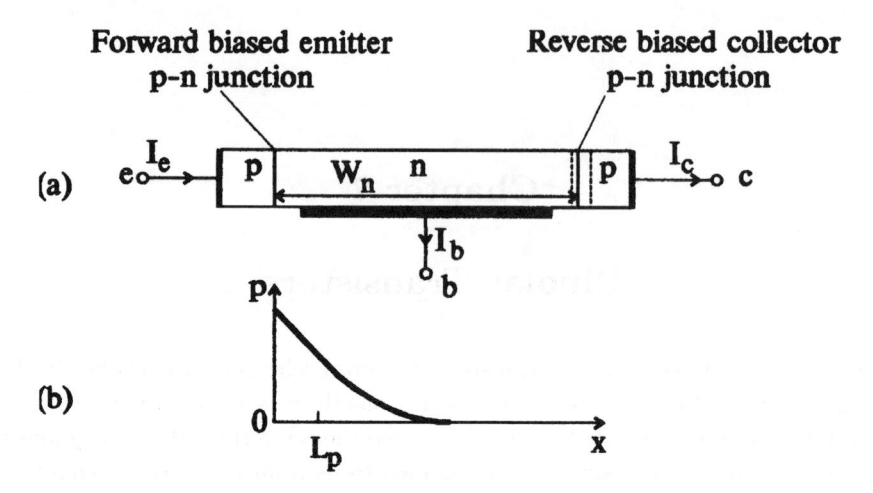

Fig. 72. Principle of operation of the structure with a "long" base. The base width W_n is much greater than the hole diffusion length L_p ($W_n/L_p \gg 1$). (a) the p-n-p structure with a "long" base; (b) the hole distribution in the base (cf. with Fig. 61). The holes which have entered the base from the emitter recombine without reaching the collector.

7.1 Principle of Operation of a Bipolar Transistor

For the transistor to be able to cope with its tasks, it is necessary that certain biases be applied to the p-n junctions forming the transistor structure.

The reverse bias must be applied to the collector p-n junction (the p-n-junction between the base and the collector). On the opposite, the forward bias should be applied to the emitter junction of a transistor (the p-n junction between the emitter and the base).

To have the necessary biases at the junctions one must connect the transistor to the external circuit containing a bias source (or sources), resistors, capacitors, etc. Such circuits may vary in types. Some of them will be described below. But now we will concentrate our attention not on the ways of creating the necessary biases, but on how the transistor acquires an ability to amplify signals, provided the necessary conditions are satisfied.

In order to have a better idea of the working principle of the transistor, let us assume, at the beginning, that the base of the transistor is not short, but long (Fig. 72). However, it is necessary to define what is meant, strictly speaking, by the terms "long" and "short". What are they compared to?

The width of the n-base W_n of the transistor is always estimated according to its ratio to the diffusion length of the holes in the base L_p. Actually in the transistors the relation W_n/L_p is always much less than unity (the base is narrow).

But we wanted first to discuss the work of the structure with a long base, in which $W_n/L_p \gg 1$ (Fig. 72). We know everything about that structure. It presents just two diodes, one of which, the left one, is forward biased, while the other, the right diode, is reverse biased. The fact that they have a common base does not affect their work at all. Nevertheless, let us again fix our attention on the working peculiarities of the junctions which are most important when analysing the transistor. A small reverse current I_c flows across the collector junction of the structure. This current is formed firstly by electrons and holes, born in the space charge region (the generation current), and secondly, by minority carriers — holes of the n-base and electrons of the p-collector which happened to come too close to the junction (the diffusion current). Let us recall again (it is very important!) that any hole born in the n-base at a distance from the p-n junction, shorter than the diffusion length of the holes L_p, has a good chance of getting within the space charge region. Should that happen, the electric field of the reverse biased collector junction will immediately toss that hole out of the base into the collector (Fig. 53).

Now let us turn our attention to the left forward biased emitter junction. The emitter of the transistor structure is always doped much stronger than the base. Therefore the mechanism of the current flowing across the emitter junction is exactly the same as that of the direct current flowing across the very asymmetric p-n junction, which had been analysed by us in detail. As we remember, the current across such a junction can be described in different ways. To analyse the work of the transistor, it is most convenient to start from the base electrode of the structure, as we had done before.

The fact that the current I_b is flowing across the base electrode of the structure means that every second $N = I_b/q$ electrons enter the base (from the external source not shown in the figure). They cannot be stored in the base in a steady state, because that state corresponds to the situation when nothing in the base changes in time, including the carrier concentration.

The electrons cannot go to the collector, because the strong field of the reverse biased collector junction forms a high potential barrier and pushes very energetically the electrons back to the base (Fig. 54). Neither can they go to the highly doped emitter (see pages 118–120). So what happens to them? They perish in the base, recombining with the holes, entering the base from the emitter through the forward biased emitter p-n junction!

The number of electrons that pass per unit time through the base electrode is the same as the number of holes which enter the base in the same period of time through the p-n junction from the emitter to the base. Speaking figuratively, one can imagine that as soon as the negatively charged electrons enter the base, they provoke the positively charged holes to leave the emitter and, diffusing across the base, kill them at the cost of their own lives.

In the structure with a long base, shown in Fig. 72(a) the holes practically fully recombine at a distance equal to several diffusion lengths L_p without reaching the collector junction. And what happens in an actual transistor structure with a short base?

Considering the processes in a real transistor structure with a short base ($W_n/L_p < 1$), it is important, first of all, to realize that the emitter and collector junctions in such a structure cannot be regarded as two isolated p-n junctions.

Why?

Because, as we have just now explained, as soon as the current appears in the base electrode of the structure, it is immediately followed by the injection of holes from the emitter into the base. On the other hand, as we know, any hole which appears in the base (its distance from the collector junction being smaller than the value L_p), has a good chance of being thrown by the field of the junction from the base into the collector. But in a transistor structure with a short base ($W_n/L_p < 1$) every hole in the base is at a distance shorter than L_p from the collector junction!

It is quite clear that part of the holes which come to the base from the emitter are sure to get to the space charge region of the collector junction. These holes will be thrown away to the collector. That means that the collector current I_c will now be determined not only (and even not so much) by the current of the reverse-biased collector junction, but also by the current of the holes flowing from the base to the collector.

The current of holes, captured by the collector, depends on the rate of the holes coming to the base, i.e. on the emitter current I_e. And that current, in its turn, depends on the magnitude of the base current I_b.

So, the currents, flowing across all the three electrodes of the structure, prove to be dependent on each other. Our task is to establish this dependence.

7.1.1 *Current amplification*

Let a structure with a short base have the base current I_b (Fig. 73). Every second $N = I_b/q$ electrons enter the base. Provoked by those electrons,

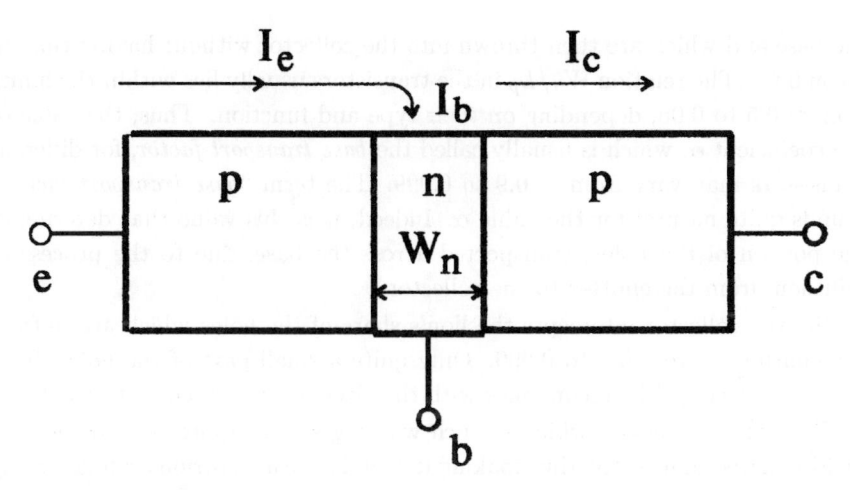

Fig. 73. The transistor structure with a "short" base. The greater part of holes passes into the collector, forming a collector current I_e. Only a small part of holes recombines in the base forming a small base current I_b.

the holes leave the emitter to kill the electrons and die. Should the base be long, the number of the electrons which might enter the base would be equal to the number of holes coming from the emitter. But in the structure with a short base it is not so. Part of the holes which enter the base will be intercepted by the field of the collector and will be thrown away from the base. They will have no time to recombine with the electrons.

Though all the electrons which enter the base must recombine by all means. They can neither leave the base, nor be stored there. That means that the number of holes which the emitter throws into the base must be greater than the number of the electrons which enter it. So, in spite of the fact that some holes will get to the collector, their number must be sufficient to provide the recombination of all the electrons that enter the base.

The thinner the base, the greater the portion of holes which enter the base, is intercepted by the collector.

If the base is short ($W_n/L_p \ll 1$), then the portion of the holes α, intercepted by the collector, is described by a simple expression

$$\alpha = 1 - W_n^2/2L_p^2 \tag{52}$$

The less W_n/L_p, the shorter the distance from the collector to the emitter junction, the greater is, naturally, the part of the holes which pass through the base and which are then thrown into the collector without having time to recombine. The relation W_n/L_p in the transistors usually lies within the limits from ≈ 0.5 to 0.05, depending on their type and function. Thus, the value of the coefficient α, which is usually called the *base transport factor*, for different transistors may vary from ≈ 0.9 to 0.999. The term "*base transport factor*" sounds quite natural for the value α. Indeed, it is this value that determines the portion of the holes, transported across the base, due to the process of diffusion, from the emitter to the collector.

So, the collector intercepts the lion's share of the holes which arrive from the emitter — from 0.9 to 0.999. Only quite a small part of the holes, from 0.1 to 0.001 (Fig. 73), recombines with the electrons which come to the base.

But this conclusion, which in such wording sounds quite pessimistic, can be formulated differently, thus making it sounds like a victorious call deserving a Nobel Prize!

The current which gets to the transistor base causes the appearance of the emitter and collector current which is tens, hundreds and even thousands of times greater.

Thus, if the current which must be amplified is applied to the transistor base, and the output signal is registered in the collector or emitter circuit, the signal will be amplified tens, hundreds and even thousands of times. The current amplification factor (*current gain*) of the transistor β is determined by the ratio of the current of the collector to that of the base: $\beta = I_c/I_b$, and, naturally, the thinner the base, i.e. the less is the value of W_n/L_p, the larger the β.

Let us now define the relation between the currents flowing across the electrodes of the transistor. It is not difficult to do it: the current of the collector I_c is, as we know, a combination of two components — the reverse current of the junction and the current of the holes coming to the collector from the emitter. In the overwhelming majority of cases, the second component is many times larger.

Then

$$I_c = \alpha I_e \tag{53}$$

On the other hand

$$I_e = I_b + I_c \tag{54}$$

Equation (54) just reflects the fact, which is quite evident to us, that the holes which have left the emitter either recombine in the base (I_b) or leave for the collector (I_c). From Eqs. (53) and (54) it follows that

$$I_c = \frac{\alpha I_b}{1 - \alpha} = \beta I_b \qquad (55)$$

where the current gain of the transistor β is equal to

$$\beta = \frac{\alpha}{1 - \alpha} \qquad (56)$$

With $\alpha = 0.9$, $\beta = 9$; $\beta = 999$ corresponds to the value $\alpha = 0.999$.

The above mechanism of amplifying the current by the transistor forms the basis of the operation of a bipolar transistor.

By the way, why do we say a bipolar transistor? As we know, the carriers of both types — electrons and holes — are equally important for the transistor operation. The electrons enter the n-base of the p-n-p transistor and cause the appearance of the hole current in both the emitter and collector, that current being much stronger.[a]

7.1.2 Parable about what is main and what is minute

Once upon a time, in a certain town, two men of wisdom taught people how they must live in this world. One sage was very strict while the other was very kind. Anyone who meditated on how one should live in this world knows that it may take (and must take) a lifetime to think about it. The disciples of those men of wisdom spent many years trying to perceive the mysteries of their teaching.

Once a stranger came to the wise man who was strict and said: "Teacher, I crave for wisdom, light and truth. I will be your most diligent pupil. I am ready to learn as long as I live. But I beg of you — render the most important things of your teaching during the time I can stand on one foot". The strict wise man sent him away. Then that stranger went to the kind wise man and asked him the same. "Stand on one foot", said the wise man grinning. And when the man stood like a heron on one foot, he said: "Do not behave towards others in the way you do not want them to behave towards you. That is the main thing. Everything else is just a commentary. Go and study!"

---

[a] In the transistor structure of the n-p-n type, whose work we hope you'll analyse yourself, the holes which enter the p-base cause the appearance of the electron current in both the emitter and collector, that current being many times stronger.

We have studied the main thing in the work of a bipolar transistor — the principle of current gain. And now we will consider some possible commentaries.

7.1.3 *Speed of response of the transistor*

The speed of response of the transistor, along with the current gain β, is one of its most important features. The speed of response of the transistor can be characterized either by the frequency limit f_c, or by the so-called rise time t_0. By rise time, we mean the following. Let the input signal be applied instantly to the transistor. For instance, the input current increases instantly by the value ΔI_{in}. If the input signal is applied for a long time, there will be a corresponding signal at the output — the output current increasing by the value ΔI_{out}. But that will not occur instantly. It will take the time t_0 that passes from the moment the input signal is applied to the moment when the output signal is established. It is the so-called *rise time* t_0. The smaller the rise time t_0, the higher the frequency f_c : $f_c \approx 1/t_0$.

Indeed, let us imagine that we have instantly applied the input signal ΔI_{in} and kept it on at the input during the time $t > t_0$ [Fig. 74(a)]. Then there is enough time for the output signal, correspondingly amplified, (ΔI_{out}) to be set at the output. [Fig. 74(b)]. After the time t, we will switch off the input signal and wait again for the time $t > t_0$ [Fig. 74(a)]. The output signal will have enough time to react to the switching off of the input signal and will disappear altogether [Fig. 74(b)]. It is clear, that should the input signal be switched on and off periodically, with the period $T \geq 2t_0$, i.e. with the frequency $f \leq 1/2t_0$, then the transistor will have time enough to amplify that signal.

Should the frequency of the signal be $f \geq 1/2t_0$ [Fig. 74(c)], the output signal will not have time enough to follow the input signal [Fig. 74(d)]. Reacting to the input signal, the output current I_{out} begins to increase. No sooner had it increased properly, than the input signal demands that the output current be diminished. As a result, with $f \geq 1/2t_0$ the output current "vibrates" with a small amplitude. The greater the frequency, the smaller the amplitude. The transistor does not have enough time to amplify a high-frequency signal whose frequency is $f \gg 1/t_0$.

The rise time of the output signal t_0 depends on the physical parameters of the transistor and on *the circuit configurations* of the transistor, i.e. on which electrode of the transistor the input signal is applied to and which one the output signal is taken from.

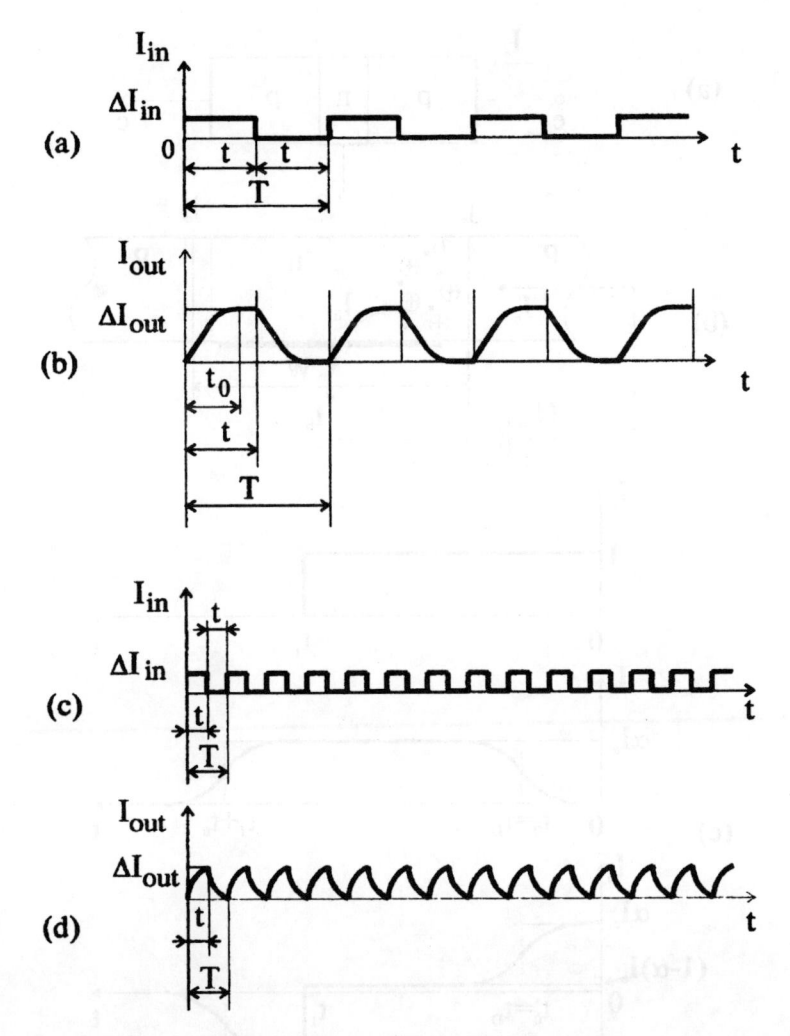

Fig. 74. If the input signal period T is much larger than the rise time t_0, then the output signal ΔI_{out} has time to reach the steady state value. (a, b). If the period T is much less than t_0, then the output signal does not have enough time to reach the steady state value (c, d).

First, let us consider what determines the rise time t_0 in the case when the input signal is applied to the emitter of the transistor and the output signal is taken from the collector (Fig. 75).

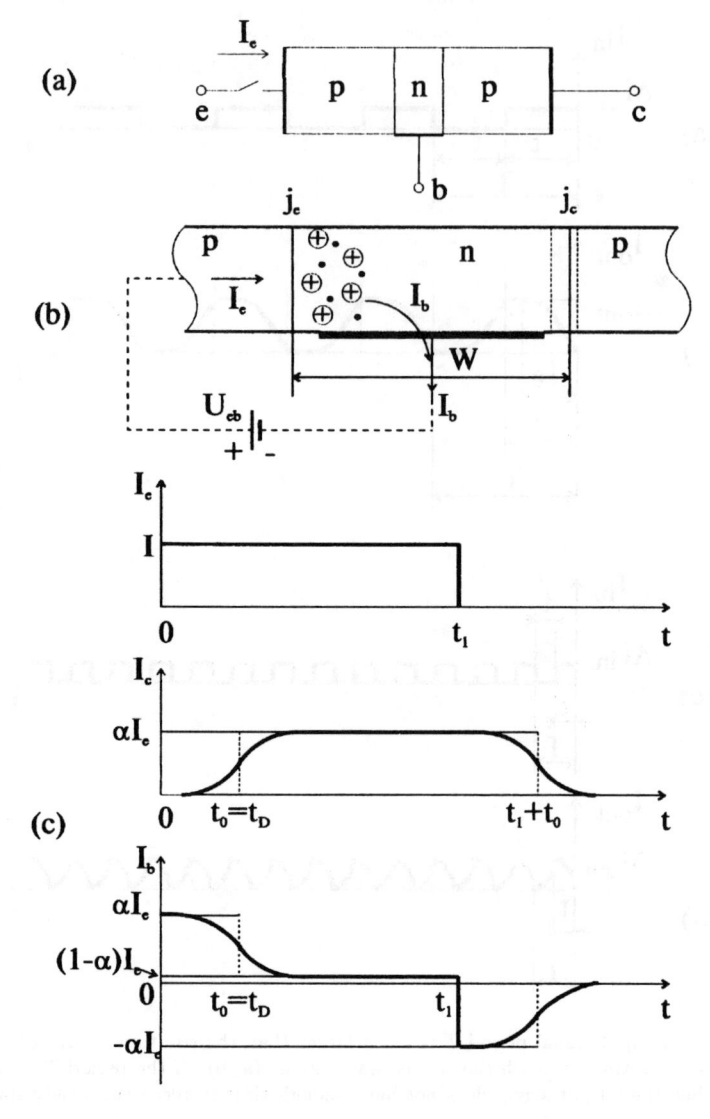

Fig. 75. Time diagram for the case when the input signal is applied to the emitter and the output signal is taken from the collector: (a) at the moment $t = 0$ the switch is closed and the holes begin to enter the base; (b) at $t < t_D \approx W^2/D_p$ the holes do not have time to reach the collector. The collector current I_c is equal to zero and $I_e = I_b$. (c) the time dependences of the current: the upper curve presents the emitter current I_e; the middle curve presents the collector current I_c; the lower curve shows the base current I_b;. Pay attention that any moment of time $I_e = I_b + I_c$;.

On the face of it, it may seem that studying the work of the transistor in that circuit is just a waste of time. What kind of gain can there be? We know too well that the current in the collector is α times smaller than that in the emitter [Eq. (53)]. Though the value α is not much different from unity ($0.9 \leq \alpha \leq 0.999$), it is anyway less than one! Any increment in the emitter current ΔI_e will cause the increment in the collector current $\Delta I_c = \alpha \cdot \Delta I_e$, the latter being smaller than the input signal ΔI_e.

Indeed, when the input signal is being applied to the emitter, there is no current gain there. But as we will see later, in this case there may be voltage gain and power gain.

We will see that with this circuit configuration of the transistor, the rise time is minimal, i.e. in this circuit configuration the transistor can be used for amplifying and generating signals of the highest frequency.

Let the switch shown in Fig. 75(a) be closed at a certain moment $t = 0$ and then the current I_e begins to flow across the emitter electrode. As we know, that means that holes begin to be injected to the base, coming across the emitter p-n junction j_e.

In a steady state the lion's portion of those holes will go to the collector, forming the collector current $I_c = \alpha \cdot I_e$. It is clear, however, that in order to be drawn into the collector, the holes, injected across the emitter, should first of all reach that collector. Since the holes move across the base due to the diffusion, it will take them the time $t_D \approx W^2/D_p$ (W being the width of the base and D_p — the hole diffusion coefficient).

What will happen during the time $0 < t < t_D$ while the holes are on their way to the collector?

Figure 75(b) helps us answer that question (that figure repeating in fact Fig. 75(a) on a larger scale and with more details). The holes injected across the emitter junction j_e had time to pass just a small part of the base and are close to the emitter junction. The collector junction is not aware of what is going on in the base. The current across the collector is then equal to zero (in case the current of the reverse biased junction is neglected). But it goes without saying that the current cannot flow only across the emitter junction. To have the current one must have a closed circuit. And before the holes reach the collector, the circuit is closed across the base electrode. The current flowing across the base I_b is equal to that flowing across the emitter I_c.

What does it mean from the point of view of physics? The holes entering the base across the emitter junction bring to the base a positive charge. Trying to charge the base positively, the holes attract the electrons from any source they may come. Obeying that "passionate call" the electrons enter the base from the base electrode. There are no obstacles there. The number of electrons entering the base per unit time from the base electrode is equal to that of holes entering the base through the emitter electrode. In short, until "the collector joins in", the current flows across the emitter junction in the same way as it flows across the forward biased p-n diode.

But after the time $t_D \approx W^2/D_p$ the holes reach the collector. Then as soon as N holes enter the base per unit time, through the emitter, the portion of holes equal to αN will leave the base for the collector. There will be the current $I_c = \alpha I_e$ in the collector electrode. Of the total number of holes N which had entered the base, there will remain only $(1-\alpha)N$ in the base. Accordingly, for the base to get charged neither positively nor negatively, $(1-\alpha)N$ electrons must now enter the base from the base electrode per unit time. Therefore the base current will be diminished to $I_b = (1-\alpha)I_e$.

The dashed lines in Fig. 75(c) show the idealized picture of the collector current I_c and the base current I_b. That picture would have been valid if all the electrons were moving across the base with the same velocity, reaching the collector at a strictly specified moment of time t_D. In that case with $t < t_D$ the current across the collector would be equal to zero. The base current I_b would be equal to the emitter current I_e. At the moment of time t_D the collector current would increase abruptly up to the value αI_e, while the base current would decrease to $(1-\alpha)I_e$.

But as we know very well, the process of diffusion is conditioned by random wandering. So part of the holes which by a happy chance has avoided any collisions in the base, will reach the collector very soon. Then those holes which are less fortunate will reach the collector later and so on. The time $t_D \approx W^2/D_p$ characterises the mean time during which the majority of the holes will reach the collector. So the dependencies $I_c(t)$ and $I_e(t)$ will not be very abrupt, they will be in fact quite blurred [the solid lines in Fig. 75(c)]. The characteristic rise time of the collector output current when the input signal is applied to the emitter proves to be of the order of time of the diffusion of holes across the base: $t_0 \approx t_D \approx W^2/D_p$.

Now let us see what will happen if at the moment t_1 the emitter current is switched off. It will become equal to zero ($I_e = 0$). And what about the collector current? That current during the very first moment will not be affected. The switching off the emitter current will instantly stop the flow of the new holes to the base. But the number of holes near the collector will remain the same. Accordingly, the number of holes getting into the collector junction j_c and forming the collector current I_c has not changed either. And (in the idealised picture) the collector current will remain unchanged, equal to αI_e until the last holes which at the moment t_1 were near the emitter junction, have diffused across the base to the collector and have been drawn in by it. It is clear that that will also take time $t_D \approx W^2/D_p$. Thus, during the switching off, the response time of the collector current is also $t_0 \approx t_D \approx W^2/D_p$.

And what will happen to the base current? One can see in Fig. 75(c) that at the moment $t = t_1$ the base current will change its direction abruptly and will grow from $I_b = (1-\alpha)I_e$ to the value αI_e. How are we to understand it? But it is quite simple. The emitter current is reduced to zero and there will be no flow of holes to the base. By that time, the concentration of electrons in the base is equal to that of holes. The base is electrically neutral. And then beginning from that moment the collector draws in the holes from the base, and no new holes enter the base. What remains in the base is just the uncompensated negative charge of "extra" electrons. This negative charge tends to push away the extra electrons and they leave for the base electrode — there being no other exit for them. The number of holes thrown by the collector per unit time is equal to the number of electrons pushed away to the base contact: $I_b = I_c = \alpha I_e$.

So, when the input signal is applied to the emitter, and the output signal is taken from the collector the characteristic time constant is $t_0 \approx W^2/D_p$.

The width of the base of the first transistors whose production began in the late forties made several tens of micrometres. With $W \approx 20$ mkm the rise time approximated $t_0 \approx t_D \approx W^2/D_p \approx 4.10^{-7}$ s, and the limiting operational frequency $f_c \approx 1/t_0$ did not exceed several hundred kilohertz. Modern super high-frequency bipolar transistors can amplify and generate signals whose frequency is tens of thousands of times greater — up to tens of gigahertz!

Let us discuss what determines the rise time t_0 in the case when the input signal is applied to the base of the transistor and the output signal is taken from the collector (Fig. 76). In that regime the transistor can amplify both the current and the voltage and that is why that circuit is used most often. Alas, everything must be accounted for. And as we will see, in this case it is the speed of response that will have to account for it. With such circuit configuration the rise time t_0 proves to be about β times greater than in the previous case that had been considered by us.

Let at the moment $t = 0$, the switch shown in Fig. 76(a) be closed, and then the base current I_b will begin to flow in the base electrode of the transistor.

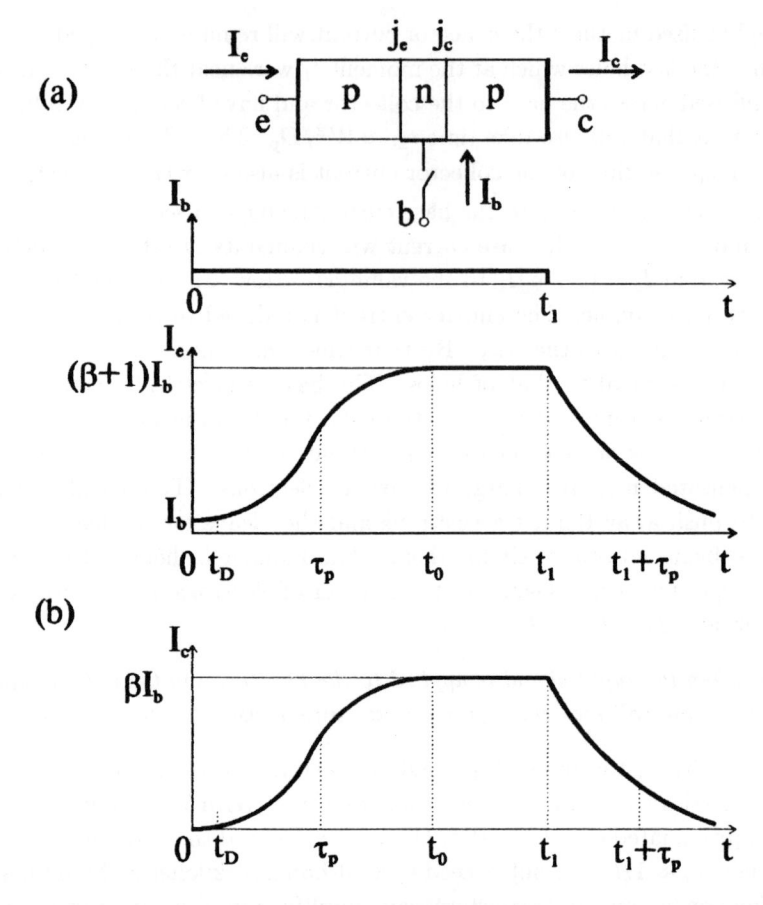

Fig. 76. Time diagram for the case when the input signal is applied to the base and the output signal is taken from the collector: (a) at the moment $t = 0$ the switch is closed and the holes begin to enter the base via the base electrode; (b) the time dependences of the current: the upper curve presents the base current I_b; the middle curve presents the emitter current I_e; the lower curve shows the collector current I_c;.

That means that $N = I_b/q$ electrons begin to enter the base. Trying to charge the base negatively, these electrons provoke the positively charged holes to leave the emitter and do their best to annihilate "the rebels". While the injected holes are reaching the collector, the number of the holes leaving the emitter must be equal to the number of electrons which enter the base: $I_e = I_b$.

It is not necessary to draw the picture similar to Fig. 75(b): it would be just an exact replica of it, the situations being the same.

We will just note that the current in the emitter I_e which flows in its circuit is $\beta + 1$ times smaller than that which will flow in the steady state when the rise time t_0 elapsed. If $\beta = 100$ (the value rather typical for transistors), the current is 101 times smaller than the one which is to be set.

But now the diffusion time of holes across the base $t_D \approx W^2/D_p$ has elapsed and the first holes injected from the emitter have reached the collector. As we remember, in the previous case, when the signal was applied to the emitter, that was the end of the transition stage. A steady state would be established [Fig. 75(c)]. But in this case, it is only from this moment that the main events begin to take place. Both the emitter and collector currents begin to grow. And as we know, the emitter current is to grow $\beta + 1$ times, i.e. ten, hundred or even thousand times! Our task is to understand what makes it grow and how long this process will last.

Let us begin our discussion with the first question. To illustrate, let us assume that 100 electrons enter the base per second. As long as $t < t_D$, one hundred holes will enter base every second from the emitter.

But after the time t_D has elapsed, the first holes reach the collector. To illustrate it, let us consider that $\alpha = 0.99(\beta = 99)$, then, beginning from that moment, from the 100 holes, coming from the emitter, 99 will flow to the collector. And only one single hole will remain in the base. The emitter calls for help: "The base sends 100 electrons per second, and from the 100 holes that I send to meet them, so that the base might be kept neutral, the collector steals 99. So now I will have to send 199 holes!"

But should the emitter send 200 holes, 198 will be stolen by the collector, and only two holes will remain in the base. If the emitter sends 300 holes, three of them will remain in the base. The emitter will always increase the number of holes it sends to the base, i.e. the emitter current and consequently the collector current will grow. How long can that go on? Until when?

Until the number of holes which remain in the base is equal to the number of electrons which enter the base (despite the fact that the collector steals 99% of the holes). It is easy to calculate that in the above example the current of the emitter will stop growing when the emitter begins to send 10000 electrons per second. It is only then the number of the holes which remain in the base will be equal to the number of electrons which enter it. So, the emitter current stops growing when relation (55) is satisfied: $I_e = (\beta + 1) \cdot I_b$. As that takes place, the current of the collector I_c will also stop growing [Fig. 76(b)] and will be

$$I_c = \alpha I_e = \beta I_b \qquad (57)$$

Let us now discuss the time it will take for the process to establish a steady state. The simplest way to do it is apart from the complex relations between the collector and the emitter, attention is fixed just on the fate of the electrons which enter the base.

As we know, the electrons which enter the base cannot leave it either for the collector or for the emitter. They are doomed to perish in the base. As soon as the number of electrons which die per unit time becomes equal to that of the electrons which enter the base, the steady state in the base is established.

The lifetime of the excess electrons in the base is equal to the lifetime of holes τ_p. And it is clear why it is so. The excess electrons attract the holes from the emitter and perish together with them. After we have recalled these familiar facts, let us now think of the time during which the steady state is being established in the base. With $t = 0$ the electrons begin to flow into the base. While $t \ll \tau_p$, practically no electrons have time to perish, and all the electrons which enter the base are accumulated there, causing the rise in the currents of both the emitter and the collector. After the time $t \approx \tau_p$ the electrons stop accumulating and a steady state is established. One can say that this assertion just follows from the definition of the value τ_p.

Indeed, while $t \ll \tau_p$, the electrons do not have time to recombine and they are stored. With t equal to several τ_p it is quite apparent that practically all the electrons which enter the base during the time $0 < t < \tau_p$ have time to die. It is therefore clear that the supply of electrons to the base at the expense of the base current is compensated by their death at the expense of recombination during the rise time $t \approx \tau_p$ [Fig. 76(b) — the central and lower curves].

So, in this circuit configuration the rise time is $t_0 \approx \tau_p$.

If at the moment t_1 when the steady state is already established, the switch in the circuit of the base is opened, the base current will be immediately reduced to zero. [$I_b = 0$, Fig. 76(b), upper curve]. The collector current I_c at first does not respond to it, the concentration of holes at the collector, determining the value I_c, at the first moment being unchanged. The emitter current I_e responds to the switching off of the base by an instant decrease of its value by the quantity $\Delta I_e = I_b$. The reason is quite clear. The number of holes sent by the emitter to the base is equal to that leaving for the collector (I_c) plus the number of holes equal to that of electrons which enter the base

(I_b). At the moment t_1, the first component (I_c) does not change, and the second component (I_b) is reduced to zero. The emitter reacts accordingly.

And then, as we can see in Fig. 76(b), both the emitter and collector currents fall down with a typical time constant which is equal to the lifetime of the holes in the base τ_p.

The thing is that after the base electrode has been opened, all the nonequilibrium excess electrons, which are stored in the base, appear to be locked there. Beginning from the moment $t = t_1$, their concentration begins to decrease at the expense of recombination. And the time constant of its decrease is, as we have just established, equal to τ_p. The collector current I_c, proportional to the hole concentration in the base, also decreases with the same time constant. And the current of the emitter I_e decreases in exactly the same way.

So, when the input signal is applied to the base and the output signal is taken from the collector, the characteristic time t_0 is equal to the lifetime of minor carriers in the base of the transistor τ_p.

It should be marked that the time τ_p is approximately β times greater than the rise time $t_0 = W^2/D_p$ in the case we had considered previously, when the signal was applied to the emitter of the transistor. Indeed, as we know, $\tau_p = L_p^2/D_p$, where L_p is the diffusion length of holes. Therefore the relation τ_p/t_0 is equal to L_p^2/W^2.

On the other hand, from Eq. (52) it follows that

$$\beta = \frac{\alpha}{1-\alpha} \approx \frac{\alpha}{1-1+W^2/2L_p^2} \approx 2\frac{L_p^2}{W^2} \tag{58}$$

When the input signal is applied to the base of the transistor and the output signal is taken from the collector, and the amplification of the transistor is at maximum, the speed of response proves to be about β times smaller than the utmost possible.

7.2 Some Words about the Types and Manufacturing of Bipolar Transistors

The industry manufactures hundreds of types of transistors: from gigantic devices, able to amplify and transform signals whose voltage is hundreds of volts and the current is tens of amperes, to tiny transistors, transforming the signals whose energy is measured in femtojouls (1 fJ = 10^{-15} J). The operation area of gigantic transistors is measured in square centimetres, while a silicon plate several square millimetres large can hold up to several million tiny transistors.

One needn't be a specialist to distinguish a powerful transistor and a tiny transistor. But one should not rely just on the size of the transistor. Similar standard packages may contain transistors with quite different parameters. On the other hand, similar transistors can have quite different packages, made either of plastic or cermet.

Let us imagine that we have carefully sawed up the package of the transistor. We see the plate of the semiconductor with three outputs — from the base, emitter and collector. We will carefully saw up the plate too and etch it in a special etchant which will make the *p*- and *n*-layers of the transistor structure visible. Having made those subtle operations with transistors of several dozen types, we will make sure that all the transistors can be divided into two large groups.

The first group will comprise the transistor structures whose design does not call for special commentaries (Fig. 77) and greatly resembles the schematic view of the transistor in Fig. 71. These are alloyed-, diffusion- or diffusion-alloyed transistors.

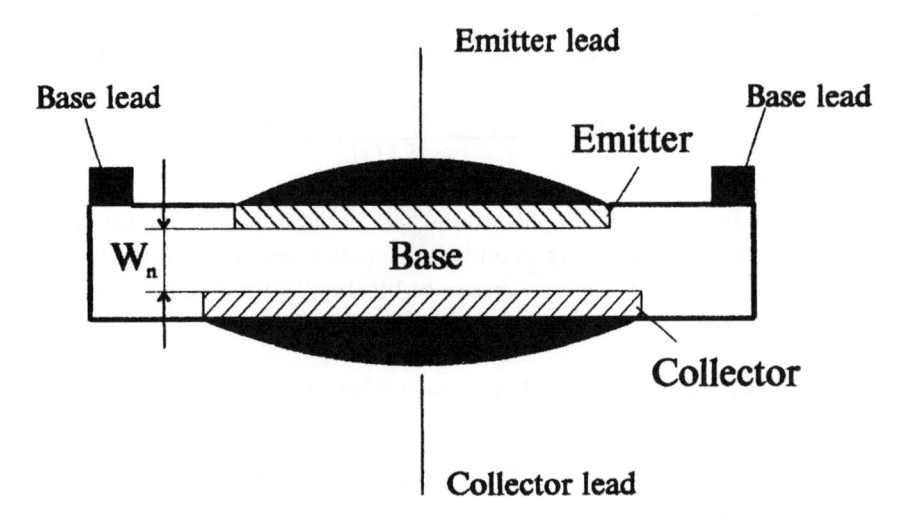

Fig. 77. Alloyed transistor.

The method of manufacturing the alloyed transistors resembles greatly the manufacture of diodes (Fig. 42). The only difference is that not one but two tablets are pressed to the initial plate (say of n-type), forming there a p-type impurity. The plate with the tablets is placed into a furnace where the alloying process takes place.

The diameter of the plate is several centimetres. The diameter of a tablet corresponds to the wanted sizes of the collector and emitter of the transistor and usually are several dozen micrometres. Therefore several dozen or even hundred tablets can be located simultaneously on the same plate. The alloying of all of them occurs simultaneously.

After alloying, the plate is cut into separate crystals — transistors, the crystals are put into a package; and the leads are bonded to the adequate electrodes. The transistors are ready (Fig. 77).

Diffusion transistors are made in a similar way. Only the diffusion is performed not from two sides, but from one side. For example to produce a p-n-p type transistor a hole semiconductor is used as the initial material. During the first diffusion a donor impurity is introduced into the crystal. As a result of that a diffusion p-n junction is formed, its depth x_1, being predetermined (compare this with Fig. 44).

Then the second diffusion is made, the acceptor impurity being introduced into the crystal on the same side. The concentration of the acceptor impurity on the surface ($x = 0$) is chosen to be greater than that of the donor impurity.

And the diffusion regime is such that the depth of the second p-n junction (point $x = x_2$) should be smaller than the depth of the first junction. As a result of that, a p-n-p structure is formed.

Apart from the technology, by which the transistors of this group are manufactured, it is specific for them that the technological operations such as alloying, applying the metal contacts, bonding of leads, etc. are performed from both sides of the initial plate.

The second group comprises the *planar transistors*. When manufacturing them, all the operations, without exception, are made from one side of the initial plate — *on one plane*. Figure 78(a) gives a schematic view of the bipolar transistor manufactured in accordance with the planar technology.

On the face of it, this design seems to be quite different from the familiar view of the transistor. It is even hard to recognize it. But nevertheless studying it attentively you can make sure that it is the same alternation of the p-n-p regions as in Figs. 71 and 77.

Figure 78(b) shows the top view of the planar transistor. One can see quite clearly the metallic strips which are the contact to the emitter (e), base (b) and collector (c) regions of the transistor.

There are many varieties of planar technology, but all of them are based on the same principle, called *selective doping*. One of the widely spread technologies of producing silicon bipolar transistors is shown in simplified form in Fig. 79.

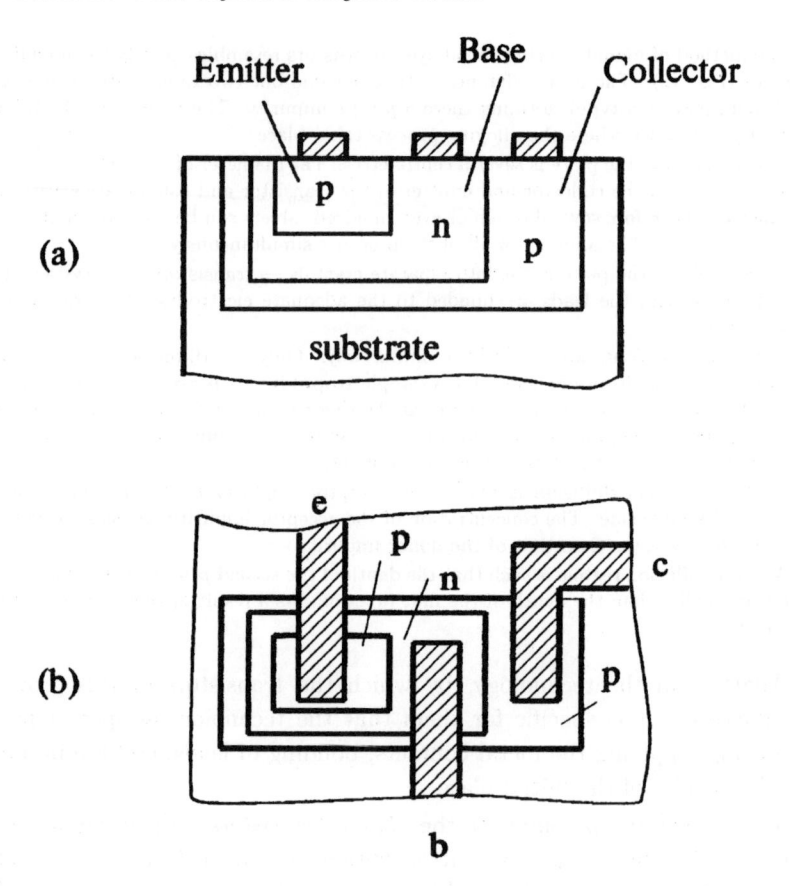

Fig. 78. The planar transistor (a) a cross-sectional view; (b) a top view.

a) The surface of the silicon plate is oxidized. A dioxide silicon layer SiO_2 is formed on it. Then the oxide is coated with a thin layer of a light-sensitive polymer — *photoresist*. The photoresist has an ability of changing its chemical properties under light. To the semiconductor plate, coated with photoresist, a glass plate is pressed. On that glass plate a picture is drawn, composed of transparent and opaque sections. After that the photoresist is illuminated through it, and those sections on which light falls acquire an ability to dissolve in certain etching agents. This operation is called *photolithography*.

b) The plate is immersed into the etchant. Those sections of the photoresist on which light fell dissolve, as well as the sections of silicon dioxide that are

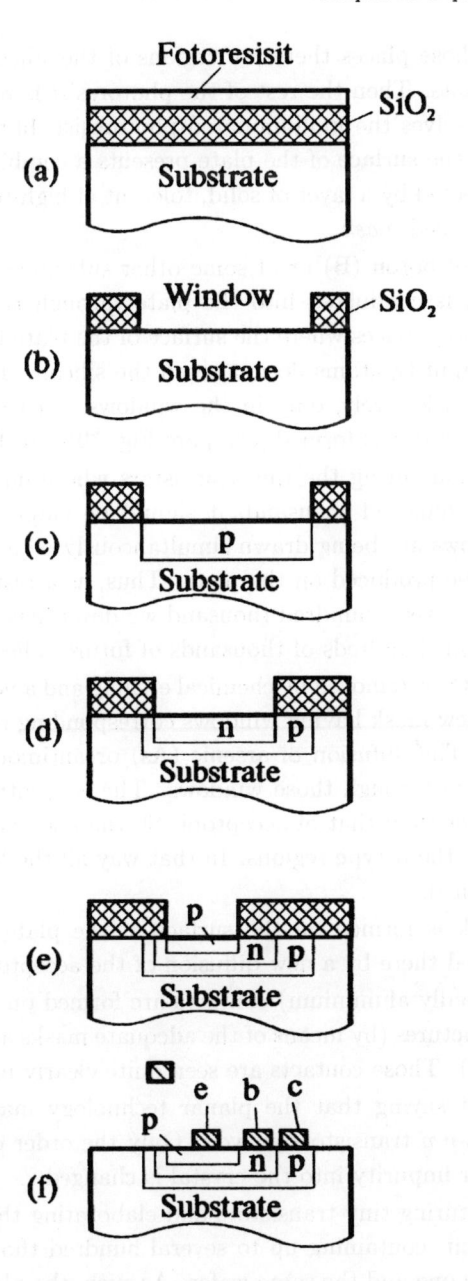

Fig. 79. The main steps of the planar transistor technology.

below them. At those places there are regions of the silicon pure surface — the so-called *windows*. Then the rest of the photoresist is removed in another etchant, which dissolves the unilluminated photoresist, but does not dissolve SiO_2. As a result, the surface of the plate presents a combination of windows and sections, protected by a layer of solid, tolerant of high temperature silicon dioxide — the so-called *mask*.

c) A diffusion of boron (B) or of some other substance, forming in Si an acceptor impurity, is conducted into the plate through the mask at a high temperature. In those places where the surface of the plate is protected by the mask SiO_2, the impurity atoms do not reach the silicon. Thus, the impurity gets to the silicon selectively, only in the windows. In that way the future collector of the transistor is formed [compare Fig. 79(c) to Fig. 78(a)].

In case of manufacturing the tiny transistors whose number on the plate may reach several hundred thousand, it should be emphasised that on the mask all the windows are being drawn simultaneously — as many of them as transistors are to be produced on the plate. Thus, as a result of one etching, there appear up to several hundred thousand windows. And as a result of one diffusion, one obtains hundreds of thousands of future collectors.

d) The mask SiO_2 is removed by chemical etching and a new mask is created on the plate, the new mask having windows corresponding to the base regions of the transistor. The diffusion of arsenic (As) or antimony (Sb) which are donors, is conducted through those windows. The concentration of donors is chosen to be greater than that of acceptors. So the regions where As and Sb are diffused will be the n-type regions. In that way all the bases of the future transistors are formed.

e) A new mask is formed on the surface of the plate and new emitter p-regions are formed there by a new diffusion of the acceptor impurity.

f) Metallic (usually aluminium) contacts are formed on the surface of the obtained p-n-p structures (by means of the adequate masks and simultaneously on the entire plate). Those contacts are seen quite clearly in Fig. 78(b).

It goes without saying that the planar technology makes it possible to manufacture the n-p-n transistors as well. Only the order of introducing the donor and acceptor impurity into the crystal is changed.

When manufacturing tiny transistors and elaborating the integral circuits (IC), a whole circuit, containing up to several hundred thousand transistors, is manufactured on one and the same wafer. As such, the planar technology is the only possible technology.

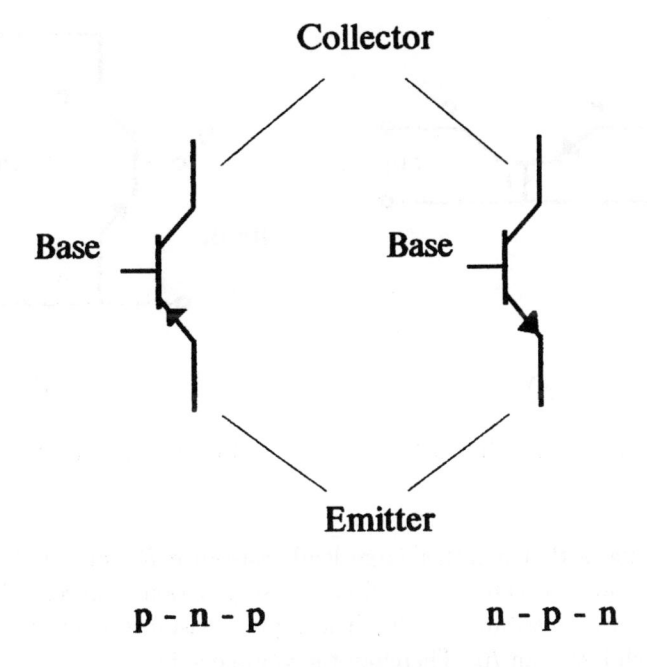

Fig. 80. Circuit symbols of bipolar transistors.

7.3 The Simplest Transistor Circuits

In this section, analyzing the simplest example, we will see how the ability of the transistors to amplify electric signals is achieved in practice.

In the circuits, the bipolar transistors are usually given in the form shown in Fig. 80. A line which meets two other lines denotes a base. The inclined line without an arrow is a collector, the inclined line with an arrow is an emitter. If the arrow is directed towards the base, that means the transistor is of a *p-n-p* type; if it is directed from the base, the transistor is of an *n-p-n* type.

If the input signal is applied to the emitter and is taken from the collector [Fig. 81(a)], it corresponds to a common base configuration. As we know, the increment of the collector current ΔI_c in this configuration is equal to $\Delta I_c = \alpha \cdot \Delta I_e$, i.e. $\Delta I_c < \Delta I_e$.

The output signal proves to be smaller than the input signal. In such a configuration the current is not gained, but gets weaker. What then do we need such a circuit?

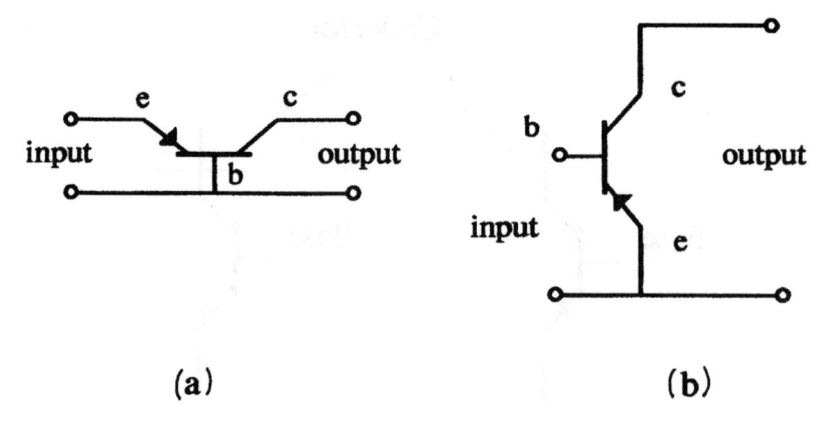

Fig. 81. The common-base (a) and common-emitter (b) circuit configurations for bipolar transistors.

The thing is that a rather large load resistance R_l can be chosen. It is possible to do it because the collector-base junction is reverse biased and presents a large resistance itself. The input resistance emitter-base R_{be} is always much less than R_l. Therefore the voltage gain

$$K_u = \frac{\Delta I_c \cdot R_l}{\Delta I_e \cdot R_{be}} = \alpha \frac{R_l}{R_{be}} \qquad (59)$$

in the common base configuration can be large enough. Besides, let us recall (see Sec. 7.1.3), that with that particular configuration the speed of response of the transistor is the greatest.

Figure 81(b) shows the common emitter circuit configuration. The input signal in that configuration is applied to the base and the output signal is taken from the collector. The current increment ΔI_b of the input signal will, as we know, correspond to the increment of the output signal $\Delta I_c = \beta \cdot \Delta I_b$. As we see, in a common emitter configuration there is a great current gain.

Besides, in this circuit configuration the voltage gain is very large, because just like it is in the common base configuration, the load resistance R_l, much greater than the input resistance R_{in}, can be chosen:

$$K_u = \frac{\Delta U_{out}}{\Delta U_{in}} = \frac{\Delta I_c \cdot R_l}{\Delta I_b \cdot R_{in}} = \beta \frac{R_l}{R_{in}} \qquad (60)$$

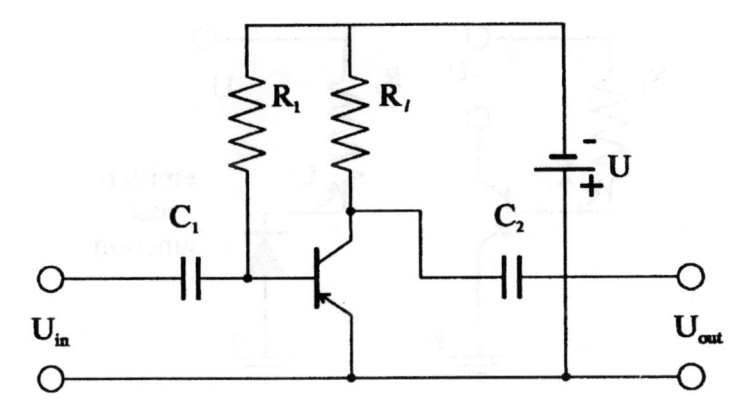

Fig. 82. The simplest BT amplifier (common-emitter configuration).

Accordingly, the common emitter circuit configuration provides a greater power gain K_p

$$K_p = K_I \cdot K_u = \beta^2 \frac{R_l}{R_{in}} \qquad (61)$$

That is why it is the most popular circuit configuration.

Figure 82 shows one of the most simple (perhaps the simplest) amplifier based on a p-n-p type bipolar transistor in a common emitter configuration. One can see that besides the transistor, the circuits comprise five more elements: the voltage source U, resistances R_1, and R_l and capacitors C_1, and C_2. What are they used for?

For the bipolar transistor to work in the regime of amplification it is necessary that its emitter junction should be forward biased, and its collector junction should be reverse biased. In the amplifier in Fig. 82, both these requirements are satisfied with only one voltage source U being used, and not two (as could be supposed). Moreover, using the voltage source U and the resistance R_1 one can prearrange the desired value of the current flowing across the transistor.

Figure 83(a) shows that the circuit: "minus" of the bias source — resistance R_1 — emitter-base junction is in fact just the resistance and the forward biased emitter p-n junction connected in series.

As we know, the voltage drop on a forward biased p-n junction cannot exceed the barrier potential difference $U_{pn} \approx E_g/q$. In actual practice, the voltage on a forward biased junction emitter-base U_{be} does not exceed

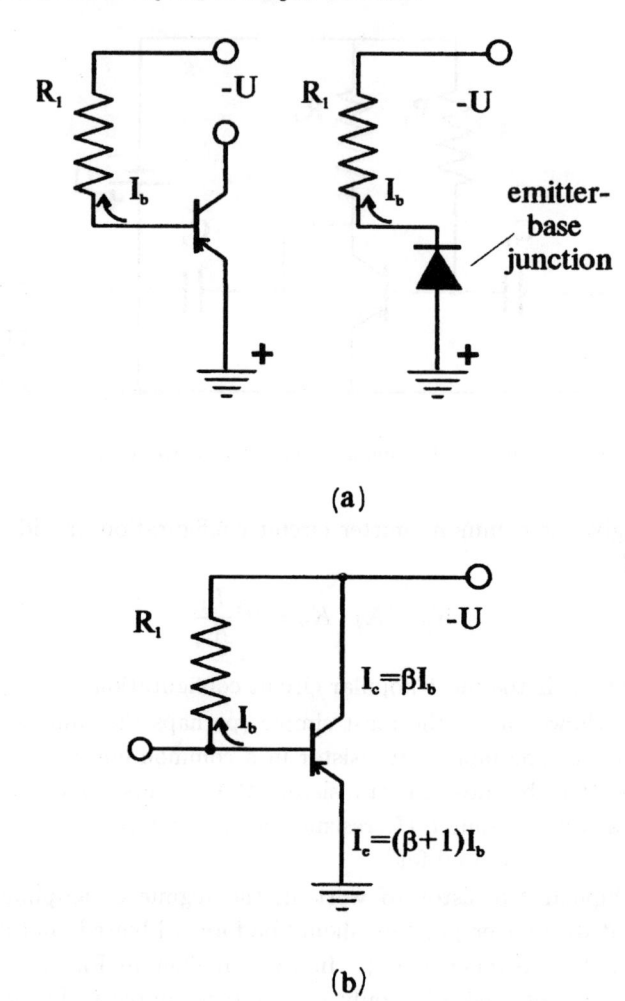

Fig. 83. The resistor R_1 determines the base current I_b (a). Consequently it also determines the emitter and the collector currents I_e and I_c (b).

the values ≈ 0.5 V for the germanium transistors, ≈ 0.8 V for those made of silicon and ≈ 1.1 V for the gallium arsenide transistors. And the source voltage U in most cases is much larger. It makes, as a rule, 3–9 V. The value of the base current of the transistor I_b can therefore be estimated quite easily [Fig. 83(a)]:

$$I_b = \frac{U - U_{\text{be}}}{R_1} \approx \frac{U}{R_1} \qquad (62)$$

The base current has been prearranged. Thus the emitter and collector currents are also prearranged and are defined by expressions (55) and (53). Changing the value of the resistance R_1, one can vary over wide limits the base current and, consequently, the currents of the emitter and collector.

The function of the resistance R_l is also clear. Should the resistance R_l be made zero ($R_l = 0$), then no changes in the input signal will affect the value of the output signal. Figure 83(b) shows that in this case the collector of the transistor will always have the constant voltage $(-U)$. With the decrease of the input signal U_{in} the value of the base current I_b will change. Any change of the base current ΔI_b will be amplified by the transistor, and the current in the collector will change by the value $\Delta I_c = \beta \cdot \Delta I_b$, but that will not affect the output voltage of the transistor when $R_l = 0$. Equations (60) and (61) make it clear that to obtain the greatest amplification factor, the load resistance should be chosen to be as large as possible. On the other hand, it is easy to understand that one should not try very hard to raise R_l — the normal transistor regime may be disturbed.

Indeed, as it is shown in Fig. 82, with the resistance R_l the potential on the collector of the transistor is equal not to $(-U)$ but to a smaller negative value: $U_c = -(U - I_c \cdot R_l)$. Should the value R_l chosen be greater than $R = U/I_c$, the current across the transistor being prearranged, at the resistance R_l all the potential of the voltage source will drop, and none of it will be left for the collector junction. The most important condition of the work of the transistor — the presence of the negative bias on the collector — will be violated. The value R_l is often chosen to be such that the potential which approximates $U/2$ should drop at the resistance R_l, the current being prearranged. When there is no input signal, then half the voltage drops at the resistance R_l and the other half drops at the transistor.

Now we have only to discuss the functions of the capacitances C_1 and C_2 in the circuit (Fig. 82). The coupling capacitor C_1 isolates dc and ac signals at the input of the transistor. Let us imagine that this capacitor does not exist in the circuit and we are using our amplifying stage to amplify the signal which is being taken from a high-quality electrodynamic microphone which is in the hands of our favorite singer. The resistance of that microphone is very small. As a result of connecting the microphone to the input of the amplifier, there will be a circuit shown in Fig. 84(a). The base of transistor will prove to be

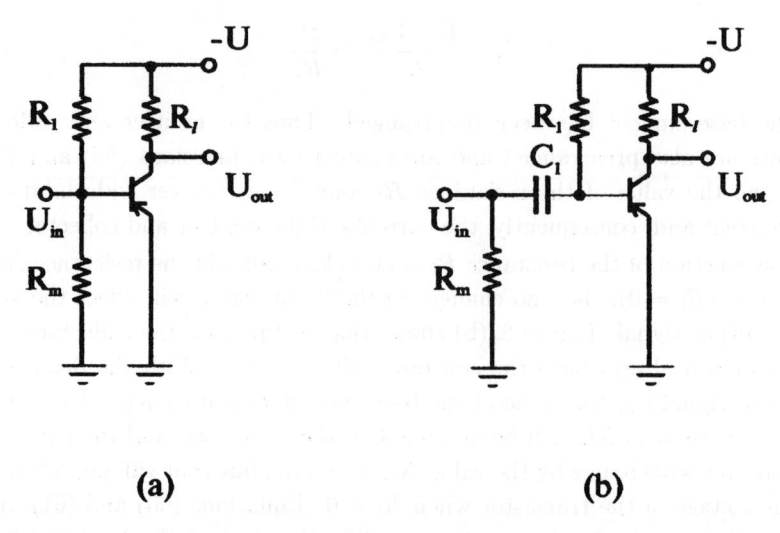

Fig. 84. Without the capacitor C_1 the connection of the low ohmic input signal source to the amplifier violates the dc regime of the transistor (a). The coupling capacitor C_1 isolates the dc and ac signals at the input of the transistor (b).

connected to the emitter across a very small resistance R_m. If we assume that $R_m = 0$, then it is clear that in this case the voltage difference between the emitter and the base will also be equal to zero (despite the presence of the resistance R_1). If $U_{be} = 0$, then the base current will also be equal to zero: $I_b = 0$, and the collector current $I_c = \beta I_b = 0$. When there is no input signal, the current across the transistor will be equal to zero. When the input signal comes from the microphone, then:

a) if this signal is negative, then it will correspond to the forward bias at the p-n junction emitter-base. Under the action of that signal there will be current in the circuit. The amplified signal will be observed at the load resistance R_l.

b) if this signal is positive, then it will correspond to the reverse bias at the base-emitter p-n junction. Such a signal tends to diminish the current flowing across the transistor. But there is nothing to be diminished, the current being equal to zero. And the transistor in such a regime does not respond to the positive signal.

As a result, the transistor operates in such a regime not as an amplifier, but as a rectifier. And instead of the familiar voice, we hear only the creak and the noise — the sound is being absolutely distorted.

Even if R_m is not equal to zero, but is small, it is said to "shunt" the input of the amplifier, changing the desired current, chosen by means of the resistance R_l. The coupling capacitance C_1 makes it possible to avoid that undesirable effect.

The value C_1 is chosen such that at the lowest possible frequency f, which must be amplified by the transistor, the capacitive resistance of the capacitor $X_c = 1/2\pi f C_1$ should be far smaller than the input resistance R_{in}. Then, for ac, the capacitor C_1 is practically a "short circuit" and whatever frequency might there be, it does not affect the work of the amplifier.

The function of the capacitor C_2 is very much the same: it isolates dc and ac signals at the output of the transistor.

We have analysed the circuit of the simplest bipolar transistor (BT) amplifier. Now, if you are interested in circuit engineering, you can take up the study of amplifiers of TVs, of radio sets, CD players, radio transmitters, etc. That is in fact an entirely different world, with its own laws, its own harmony, its criteria of usefulness, beauty and expedience.

It goes without saying that what we have analysed here are not the most modern or the most effective BT circuit, or those of the greatest temperature stability, etc. There are much more perfect circuits. They contain much more sophisticated ideas. Their methods of calculation are most elaborate. However, all of them are based on the same physical fundamentals we have studied here, analysing the most simple example.

7.4 Summary

The three-layer semiconductor structure either of the *p-n-p* type or of the *n-p-n* type under certain conditions can amplify electric signals. Thus, it is first of all necessary that the central part of the structure — the base — should be short enough. The width of the base must be shorter than the diffusion length of the minor carriers in it. The thinner the base, the greater the amplification. Besides, the thinner the base, the higher can be the frequency which the bipolar transistor (the *p-n-p* or the *n-p-n* structure) is able to amplify and generate. Modern bipolar transistors amplify the power of the signal, applied to them, tens of thousands of times and they can operate at frequencies of up to tens of billions of hertz.

Chapter 8

Field Effect Transistors

No gain without pain.

Kozma Prutkov (1803–1863)

About a quarter of a century before the first bipolar transistor was invented, in the 1920s, American researcher Julius Lilienfeld put forward a very simple, ingenious and compelling idea. They proposed the design of a semiconductor device capable of amplifying electric signals.

8.1 The Beginning

8.1.1 *The main idea*

The main idea of that device is illustrated in Fig. 85. The device is an ordinary plain capacitor [Fig. 85(a)]. One of the the capacitor plates is a layer of metal, the other is a semiconductor plate. If the voltage is applied between the plates of that capacitor, an electric field F_1 will arise in the insulating gap. On the surface of the semiconductor plate, on its interface with the insulator, an electric field $F_2 = F_1/\varepsilon$ will appear, where ε is the semiconductor dielectric constant (See Fig. 37).

We know very well (see Chapter 4) that the field, arising on the border of the plate, penetrates to a certain depth of the semiconductor. This depth depends on the free carrier concentration in it. Depending on the direction, the field may either create a depletion layer, driving the carriers away from the semiconductor surface, or it may enrich the surface layer with the excess charge carriers.

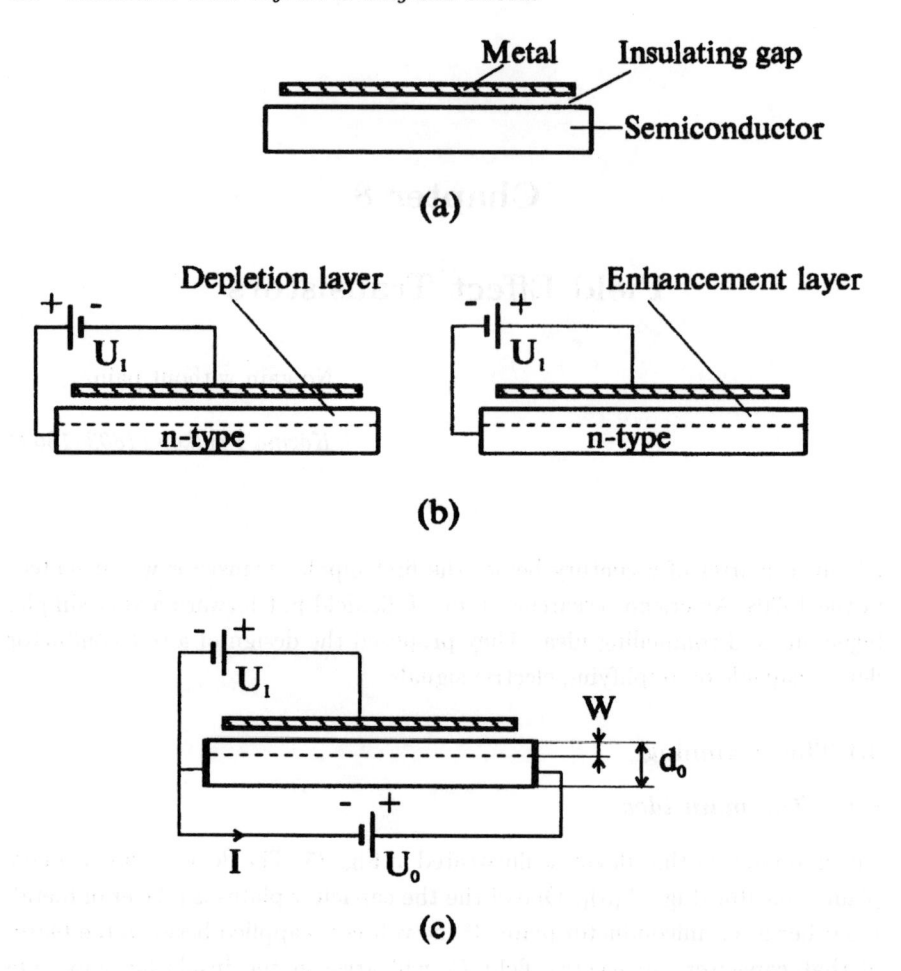

Fig. 85. The idea of the device which was put forward by Julius Lilienfeld. (a) A plain capacitor. The upper plate is a metal layer; the lower plate is a semiconductor layer. (b) If the capacitor plate is made of an n-type semiconductor, then if the potential on the metal plate is negative, the electrons will be driven out of the surface layer of the semiconductor (left). Should the polarity of the applied voltage be changed, near-surface layer of the semiconductor plate will get enriched in excess electrons (right). (c) The current I flowing in the semiconductor plate, is controlled by an electric field, perpendicular to the current.

If the capacitor plate is made of an n-type semiconductor, as is shown in Fig. 85(c), then if the potential on the metal plate is negative, the electrons will be driven out of the surface layer of the semiconductor. Should the polarity of the voltage applied to the capacitor be changed, the near-surface layer of the semiconductor plate will, on the contrary, get enriched in excess electrons.

Let us now apply the voltage U_0 along the semiconductor plate [Fig. 85(c)]. If the voltage U_1 applied to the capacitor has such polarity that part of the semiconductor plate is depleted of carriers, then the resistance of the plate increases and the current I_1, flowing along the plate is, naturally, decreased. And vice versa, should part of the plate get enriched in free carriers its resistance will naturally fall and the current will grow.

The current flowing in the semiconductor plate is controlled by an electric field perpendicular to the current.

The idea is simple, clear and fine, isn't it?

Yes, no doubt it is a brilliant idea. It has one drawback. It "doesn't work". Numerous attempts made by Lilienfeld and his followers were fruitless.

They did not manage to get any amplification or even any noticeable change in the resistance of the semiconductor plate.

8.1.2 Simple estimations

But what could Lilienfeld and his followers have relied on in the 1920s?

Suppose the polarity of the voltage U_1 applied to a semiconductor capacitor is such that the electric field which appears in the semiconductor drives off the carrier, forming a depleted layer near the surface. Then, knowing the surface field F_2 and the doping impurity concentration N_d, we can determine the width W of the depleted layer, using Eq. (31):

$$W = \frac{\varepsilon \, \varepsilon_0 F_2}{q N_d} \tag{63}$$

[See also Fig. 40(b)].

The maximal field that can be reached in the semiconductor, is the field of impact ionization F_i. For Ge and Si, the value F_i is, as we know, $(2\text{-}3) \cdot 10^5$ V/cm (see p. 127). As for the value N_d, its minimal possible value is defined by the level of technology that has been achieved. In the 1920s, the value $N_d \approx 10^{18}$cm^{-3} was an object of pride of the researchers. And just as it should be. For it corresponds to the impurity concentration in the semiconductor $\leq 0.01\%$! With $F_2 = 3 \cdot 10^7$ V/m, $N_d = 10^{24}$ m^{-3} and $\varepsilon \approx 10$, the width W of the depleted region is $\approx 2 \cdot 10^{-8}$ m, i.e. as little as 0.02 μm. On the other hand, in those years it was practically impossible to produce a semiconductor plate less than ≈ 50 μm thick. If in a plate 50 μm thick ($d_0 = 50$ μm), a layer

$W \approx 0.02$ μm thick is depleted by the carriers [Fig. 85(c)], then its resistance will increase as little as $W/d_0 = 1/2500$, i.e. by 0.04%. The calculated value is too small to practically realise the amplification effect. But one could measure very well the change in resistance of 0.04% even in the 1920s.

But the change in resistance was not observed at all in the experiments. Or it proved to be hundreds or thousands of times less than the calculated value.

8.1.3 Old acquaintances

Should the change in resistance coincide with the calculated value, there would be no grounds for pessimism. It was clear to everybody that with the development of technology, the level of purification of the semiconductor crystal would improve and the value N_d could be decreased to a great extent. With the decrease of N_d, the width of the layer W will increase. On the other hand, it was clear that the thickness of the plate d_0 could also be diminished.

But alas, the measured value proved to be much less than it had been expected. And that made the validity of the idea and the possibility of its realization rather doubtful.

The researchers repeated the experiments again and again. They changed the geometry of the capacitor, the material of the insulating gap, the type and character of doping of the semiconductor, the temperature of the experiment — everything was in vain.

Meanwhile, the explanation lay literally on the surface. It was our old acquaintances — *the surface states* (Section 3.2) — that were responsible for those negative and discouraging results of the experiments.

If every atom that is on the surface creates a surface level, and that is what unfortunately happens most often, then the density of the surface states is, as we know, $\approx 10^{15}$ cm^{-2}. Each of those surface levels is willing to capture a free carrier, should it happen to be near the surface.

Now imagine that the field F_2 that appeared on the surface of the semiconductor under the controlling voltage U_1 draws the excess electrons up to the surface. If there were no surface states, those excess electrons would increase the free carrier concentration in the near-surface region. As a result, the resistance of the near-surface region, and consequently of the whole semiconductor plate would be much smaller. However, if the surface states density is greater than the concentration of the excess electrons drawn to the surface, then all the carriers will be captured at the surface levels and, consequently, will lose their ability to move under the action of the field. These motionless electrons, bound on the surface states, will not affect at all the resistance of the near-surface layer. Now we have to compare the density of surface states to the surface density of those carriers which can be drawn to the surface (or away from the surface) by the field F_2.

The relation between the surface charge density σ and the field F is established by the expression $\sigma = \varepsilon\varepsilon_0 F$ (see Section 4.3).

With $\varepsilon = 10$ and $F = 3 \cdot 10^7$ V/m the value of σ is $2.5 \cdot 10^{-3}$ C/m^2. Dividing this value by the charge of electron $q = 1.6 \cdot 10^{-19}$ C, we obtain the surface density of carriers. As it is not hard to calculate, it is $\approx 1.5 \cdot 10^{16}$ m^{-2} or $\approx 10^{12}$ cm^{-2} — 1000 times less than the value of the typical density of surface states on the real surface! It is clear why the researchers had failed to achieve any results for so many years.

8.2 Maturity and Flourishing

The invention of a bipolar transistor in 1948 did not stop any attempts to create a device the idea of which we have been discussing.

The first man who succeeded in creating the *field effect transistor* (that was the name given to this hypothetical device after the bipolar transistor had been invented) was William Shockley — one of the creators of the bipolar device.

8.2.1 *JFET (p-n-junction field effect transistor)*

The idea suggested by Shockley in 1952 was ingenious, brave and quite unexpected. How did the researchers reason before? It was difficult to produce the field effect transistor due to the surface states, wasn't it? Then it was necessary to try and get rid of them. They would try one, two, three semiconductors .... They would use quite a number of materials for the insulating gap .... They would change the technology of deposition of the dielectric layers. They would try sputtering, chemical deposition, changing the temperature .... It was all to no avail.

How did Shockley reason? It is because of the surface states, isnt it? Let them be. The same for the surface! Let us create an electrode which will modulate the resistance of the semiconductor plate not near the surface, but in the bulk, where there are no surface states.

Figure 86 illustrates Shockley's idea. In a thin n-type plate, a p-type region is created on top of it. It can be done using any of the methods known to us

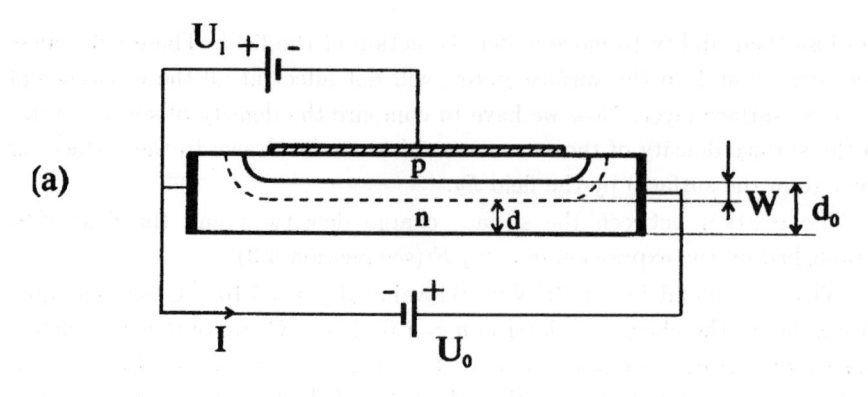

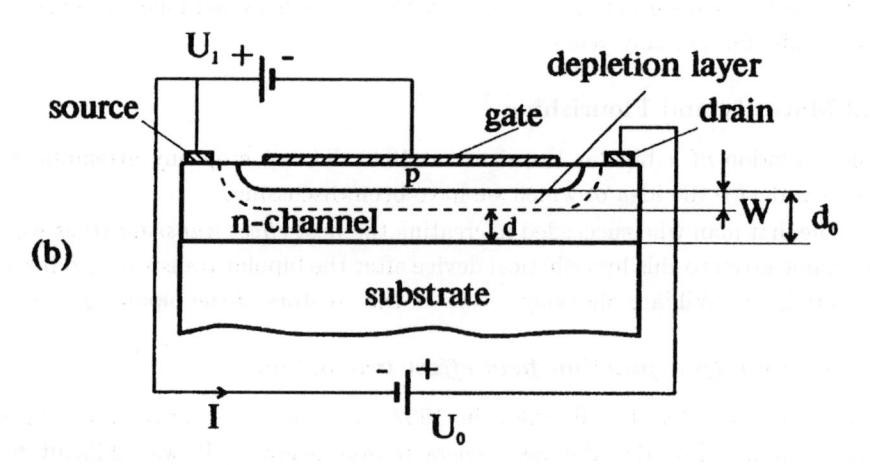

Fig. 86. Field effect transistor with the *p-n* junction (**JFET**). (a) Schematic design; (b) JFET with an *n*-channel.

(by diffusion, by alloying, by ion implantation). In the depth of the plate, at a desired and strictly regulated distance from the surface, there appears a *p-n*-junction. The *p*-type region is doped much stronger than the *n*-region. Therefore the depleted region of the *p-n*-junction is mainly located in the *n*-layer.

If the reverse bias U_1 is applied to the *p-n*-junction, then the depletion layer, which has a very large resistance (*the space charge layer*), will with the growth of the bias penetrate into the depth of the *n*-region. The channel, along which the current I may flow, is getting narrower, while the resistance of that channel is always growing. With the voltage U_0 being constant, the larger the value of the reverse bias U_1, the weaker is the current I.

The current flowing along the semiconductor plate, is regulated by the electric field which is perpendicular to the current! [compare this to Fig. 85(c)]. Figure 86(b) shows not the principal, but the real design of the field effect transistor with a p-n-junction. The transistor is manufactured according to the planar technology. All the three electrodes are on one side of the semiconductor plate, on top of it. The electrodes which serve to pass the current along the channel are called *source* and *drain*. The electrode to which the voltage modulating the resistance of the channel is applied is called the *gate*.

The electrode through which the carrier enters the channel is called the source. Figure 86(b) shows a field effect transistor (FET) with the n-channel. The carriers (electrons) enter the channel from the electrode to which the "minus" of the battery U_o is applied. Transistors with the p-type channel are also produced. In this case the "plus" of the battery U_0 is applied to the source, and the "minus", to the drain. The voltage, modulating the resistance of the channel is applied, as a rule, between the source and the gate, and it is called the gate voltage U_g.

8.2.2 *Fortune favours the brave. MOSFET*

A few years after the creation of a field effect transistor with a p-n junction, at the end of the 1950s, the patience and skill of technologists enabled them to win a victory in the struggle which lasted for more than 30 years.

At last they found such a semiconductor, such a material of the dielectric gap between the metal- and semiconductor plates, and such a method of applying a dielectric layer on the semiconductor which might provide the surface state density not exceeding $\approx 10^{10}$ cm^{-2}, i.e. 100 000 times smaller than the typical value of the semiconductor surface state density.

It is very instructive (and most important) that it was ... silicon which appeared to be that very semiconductor that provided the success of the operation. As for the dielectric, it proved to be silicon oxide, SiO_2.

And now after hundreds of thousands of unsuccessful experiments, the process of creating the field effect transistors which fully comply with Lilienfelds, idea seems to be quite simple.

A thin silicon layer with the necessary type of doping and concentration — the future channel — is grown on the substrate. Then the surface of the

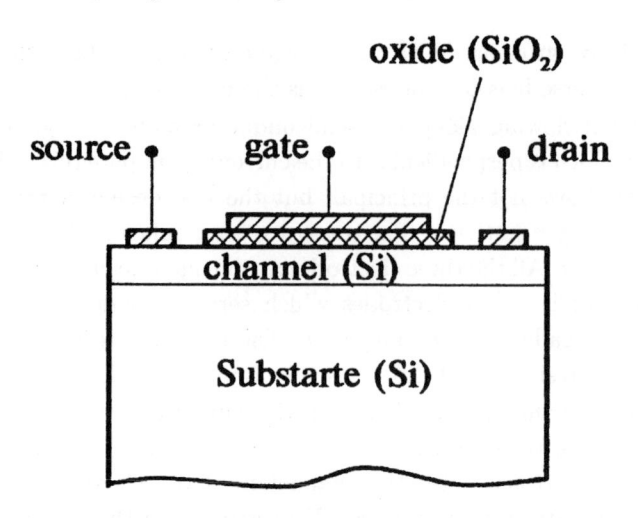

Fig. 87. Schematic diagram of **MOSFET**.

silicon is oxidized. The so-called "windows" are etched in the oxide by means of the process of photolithography, already familiar to us (p. 188).

A deposition of the metal is conducted in those windows forming source and drain. And at the place of the gate, where the oxide remained untapped, metal is applied on the top of the oxide, and the transistor is ready (Fig. 87). Such transistors are called MOSFET — Metal-Oxide-Semiconductor FET.

When producing MOSFETs, it is often convenient to use as a substrate not a semi-insulating material but a comparatively low-doped semiconductor whose conduction type is opposite to that of the channel, i.e. to create a transistor with an n-channel, it is necessary to use a low-doped p-type silicon (and for the transistors with a p-channel, the substrate of n-type).

Figure 88 shows the two most important types of the MOS transistors: the transistor with a *built-in channel* [it is also called *depletion mode or normally-on* FET, Fig. 88(a)] and that with the *induced channel*, [it is also called *enhancement mode or normally-off* FET, Fig. 88(b)]. Both of the them, shown in Fig. 88, are transistors with the n-type channels and they are made on the p-type substrate. The MOSFETs with the p-channel on the substrates of the n-type are manufactured in the same way.

The transistor with the *built-in channel*, shown in Fig. 88(a), is very similar to the transistors shown in Figs. 85 and 86 by its design and its principle of

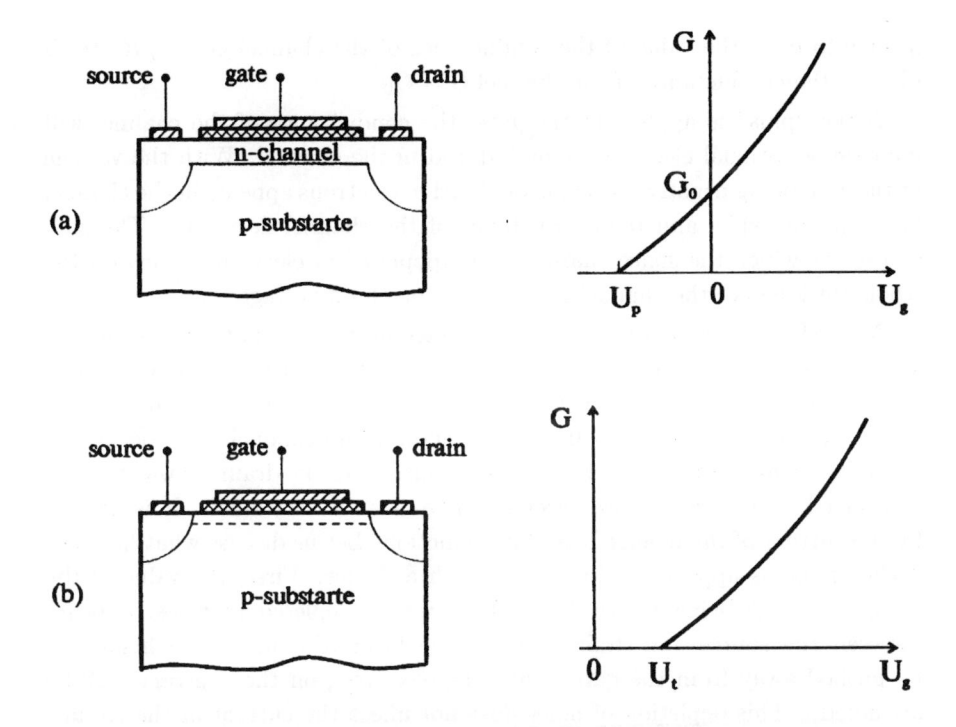

Fig. 88. Different types of MOSFETs (compare to Fig. 87).

operation. Strictly speaking, the difference lies only in the design of the source and drain. As is seen in the picture, to form the drain and source regions, before the deposition of metal contacts, two deep p-n-junctions are formed (by means of diffusion and ion implantation) in the p-type substrate. The "deep" n-regions serve as reliable and good quality contacts to the n-channel. And the current between these n-regions can practically flow only along the n-channel. Indeed, the circuit: source-substrate-drain presents in fact two p-n junctions connected in series. No matter what the voltage polarity applied between the drain and the source might be, one of those p-n-junctions will always be reverse biased, and the current in the circuit source-substrate-drain will prove to be negligible. When no voltage is applied to the gate, $(U_g = 0)$ the MOSFET is in the "on" state. In this case the transistor is called a normally on-channel transistor. The resistance of the channel R_0 with $U_g = 0$ is defined by the length of the channel L, and by the thickness and conductivity of the channel. Speaking of FETs, they often prefer to speak not of resistance, but of the value inverse to it — the value of the conductance of the channel $G = 1/R$. With $U_g = 0$, the conductance of the channel $G = G_0$.

If the "plus" is applied to the gate, the conductance of the channel will increase, additional electrons being induced in the channel. With the voltage at the gate being negative, a region depleted of electrons appears in the channel beneath the oxide, and the conductance of the channel decreases. The gate voltage at which the space charge region, depleted of electrons, pinch off the whole thickness of the channel is called the pinch off voltage U_p.

MOSFET shown in Fig. 88(b) on the face of it looks quite extraordinary. There is no special channel in this transistor between the source and drain. It is a field effect transistor with a *normally-off* channel. When no voltage is applied to the gate, ($U_g = 0$) there is actually no channel whatsoever. At any polarity and any voltage between the source and the drain, as was already mentioned, the current in the circuit source-drain is very small. It is defined by the current of the reverse biased p-n junction. Let us discuss what happens if the "plus" is applied to the gate of such a device. First, the value of the applied voltage being relatively small, a region depleted of holes, appears between the source and drain beneath the layer of oxide. The holes will be pushed away from the gate (while the electrons, on the contrary, will be attracted). This depletion of holes does not affect the current in the circuit-source-drain. It will be still practically equal to zero. But should the positive voltage at the gate U_g be large enough, there will be a phenomenon, familiar to that in Chapter 3 — the inversion of the type of conductivity (Section 3.3). An inverse n-type layer will appear beneath the gate, which will create an n-channel between the source and the drain.

A sufficiently large voltage at the gate induces a channel for the current between the source and drain. The gate voltage at which the inversion takes place, forming a channel, is called the *threshold voltage* and is usually designated by the symbol U_t. With $U_g > U_t$ the conductance of the channel grows with the increase of U_g, since with the growth of the positive bias at the gate, more and more electrons are attracted to the surface, their concentration in the channel increasing.

The main advantage of the FETs with the induced (normally-off) channel, providing their wide usage, is the circumstance that if there is no voltage at the gate, they practically do not conduct any current and, consequently, they do not consume any power even in the case when the voltage between source and drain is applied.

8.3 Epitaxy

Before we begin to discuss the FETs operation regimes we'll briefly speak of producing very thin semiconductor films, of decimal or even hundredth fractions of micron thick. Those films are quite essential for the work of modern semiconductor devices: the FETs and BTs, varicaps, photo-diodes and LEDs, etc. As a rule such films are grown on comparatively thick substrates. Several methods, called the *epitaxial methods*, have been devised in order to grow films on substrates. The term epitaxy is a combination of two Greek words: *"taxis"* — (arranged in order) and *"epi"* — "on", "on top of". It is a very expressive term — it reminds us that the monocrystal (ordered) layer of material is grown on top of the substrate.

There are 3 main types of epitaxy: the vapor phase epitaxy (VPE), the liquid phase (LPE) and the molecular-beam epitaxy (MBE).

In case of a *vapor phase epitaxy*, atoms of the material are deposited on the substrate from a flow of gas. The VPE scheme is shown in Fig. 89. The quartz tube (1) contains a cassette (2) with a number of substrates (3). The atoms of the material are deposited on the substrate heated to a certain temperature from a flow of gas (4) passing above the cassette.

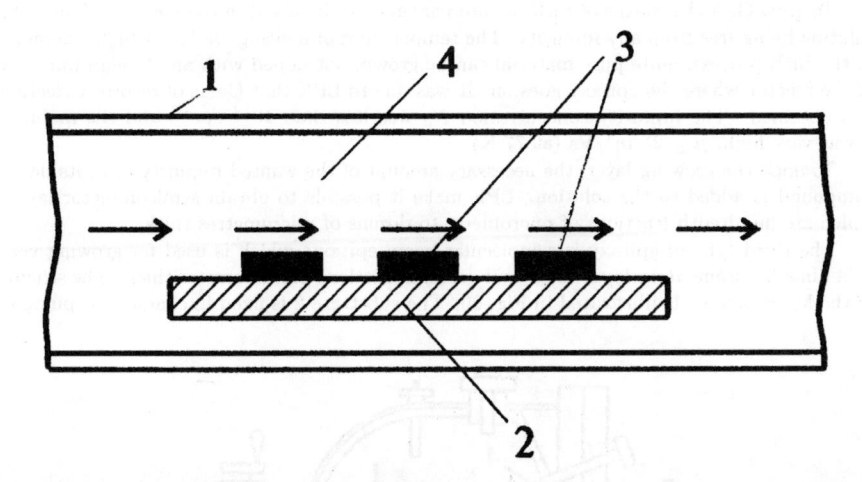

Fig. 89. Schematic view of the vapor phase epitaxy (VPE). 1 — quartz tube; 2 — cassette with a number of substrates 3; 4 — flow of gas.

In case of the epitaxy of silicon, a reaction of reduction of the silicon tetrachloride is widely used: $SiCl_4 + 2H_2 \rightarrow Si + 4HCl\uparrow$. Vapours of $SiCl_4$ and of hydrogen are supplied to the reactor. The silicon atoms, formed as a result of the reaction, are deposited on the substrate which has a temperature of $1150-1200°C$. These atoms wander over the substarte surface a certain time until they find a suitable place and establish good connections with the atoms of the substrate. An epitaxial film begins to grow on the substrate. The rate of growth of the film in the VPE process makes about 1 μm per minute. The layers from 1 micrometre to several dozen micrometres thick are grown by means of VPE.

It is very important that the process of epitaxy proceeds at a temperature significantly lower than the temperature of melting at which crystals are usually grown. So, the temperature of melting silicon makes 1417°C and the temperature of the epitaxial growth is by almost 300°C lower. A lower temperature results in the fact that the growing film is far less doped with foreign impurities. The semiconductor materials of a record perfection are obtained, as a rule, by epitaxial methods.

In the process of epitaxy it is possible to regulate the doping of the growing layer. During the vapour phase epitaxy in order to dope the growing film, vapours of the substance containing the doping impurity are introduced to the reactor.

VPE epitaxy is used to grow the layers not only of silicon but also of germanium, gallium arsenide and many other semiconductor materials.

While LPE is used, the atoms of the growing layer are deposited on the substrate from the melt or solution. The solution (melt) saturated with the semiconductor material which is to be grown is applied to the substrate. Then both the substrate and the melt are cooled. With the decrease of temperature the solubility of the material is lowered and the excess quantity of the semiconductor is deposited on the substrate. In case of LPE silicon process the solutions of Al-Si and Au-Si are used.

To grow GaAs by means of LPE, gallium arsenide is dissolved in the melted gallium, the solution being free from any impurity. The temperature of melting Ga is not high, therefore in the LPE process, quite pure material can be grown, not doped with any foreign impurity of the reactor where the epitaxy goes on. It was due to LPE that GaAs of record perfection was obtained. The impurity concentration N_d was less than 10^{12} cm^{-3} and the mobility μ was very high: $\mu \geq 25$ m^2/Vs (at 77 K).

To dope the growing layer, the necessary amount of the wanted impurity or of its liquid compound is added to the solution. LPE make it possible to obtain semiconductor layers which are hundredth fractions of micrometre to dozens of micrometres thick.

The third type of epitaxy is a molecular-beam epitaxy, which is used for growing very thin films from *one atom layer* to several decimal fractions of micrometre thick. The scheme of the MBE process is illustrated in Fig. 90. The substrate 2 is fixed in chamber 1, pumped

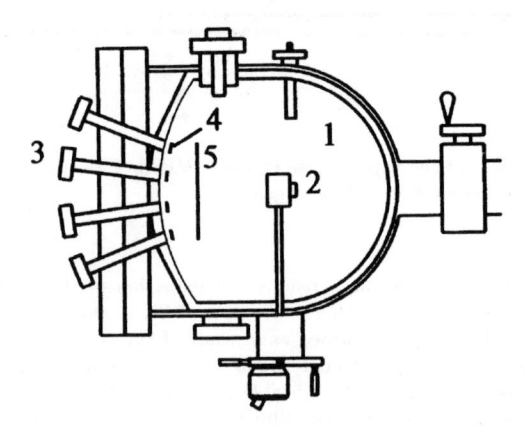

Fig. 90. Schematic view of the molecular beam epitaxy (MBE). 1 — chamber; 2 — substrate; 3 — evaporation sources; 4 — shutters; 5 — main shutter.

to a very low pressure. A few evaporation sources 3 are located in the same chamber in a semicircle around the substrate. Each source is separated from the main space of the chamber by a shutter 4 which can open and close very rapidly, obeying the command of the computer. Every source contains a substance which will be required in the process of epitaxy, and also a heater. The heater maintains the temperature, necessary to evaporate intensively the adequate material from the source. Thus, in every source there is a sufficiently high excess pressure of vapours of the material, contained in the source. By the command of the computer, the shutter which separates the source from the chamber is off and a beam of molecules rushes from the source into the chamber, flies to the substrate and is deposited on it, getting crystallised.

Several shutters can be opened simultaneously, then molecules of different sorts can get to the substrate.

By means of the process of a molecular-beam epitaxy it is possible to grow very thin layers of complicated semiconductor compounds, strictly regulating the thickness, composition and concentration of impurities in every layer (Fig. 91).

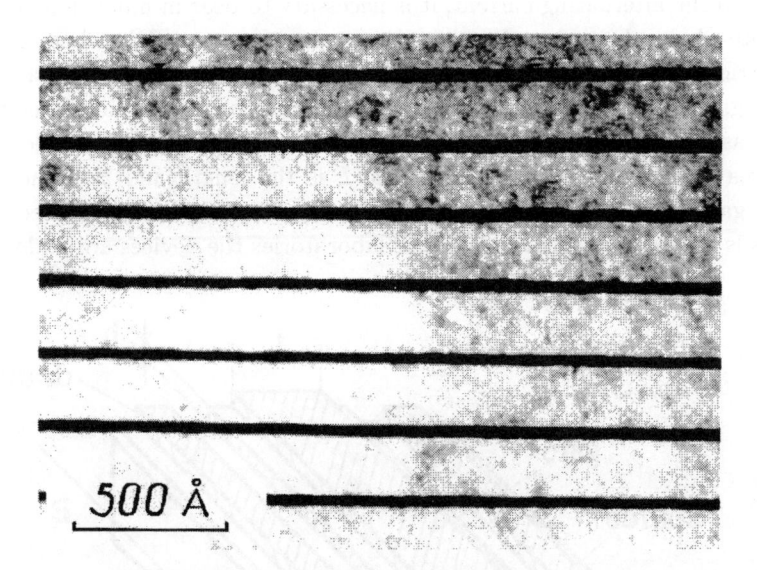

Fig. 91. The multilayer InGaP/InGaAsP strucutre. Dark strips are the InGaAsP layers; the layer's thickness is 40 A. Light strips are the InGaP layers; the layer's thickness is 180 A. Courtesy by Prof. N. Berdt, Ioffe Institute, St. Petersburg, Russia.

8.4 A Few Important Details

One and the same property — the ability to amplify signals — is realised in the bipolar and field effect transistors based on quite different principles.

In BT the change of the output current in the circuit emitter-collector is caused by the change of the input base current. The input current flows across the forward-biased p-n-junction. Therefore the input resistance of the bipolar transistor is not large — as a rule it lies within the limits of a few ohms to several kiloohms.

In the FET the change of the output current in the circuit source-drain is caused by the change in the gate voltage. The input voltage U_g is applied to a very large resistance of a reverse-biased p-n junction or to a still larger resistance of the isolating plate (SiO_2). Therefore the input resistance of the FET of any type is, as a rule, very large. It is millions and in some cases thousands of millions of ohms (gigaohms).

Such enormous resistance exists in the input circuit of the FET for the direct current. In order to estimate the input resistance of the FET with regard to the alternating current, it is necessary to bear in mind that the gate is, quite obviously, a capacitance, whose value $C_g = \varepsilon\varepsilon_0 S/h$, where ε is the dielectric constant of the isolating layer, S is the gates area, h is the thickness of the gate insulator. In MOSFET the thickness h of the oxide layer which serves as an insulator is $\approx 0.1\mu m$, the dielectric constant of the oxide $\varepsilon = 4$, the area of the gate is $S = LB$, where L is the length and B is the width of the gate (Fig. 92). The length of the gate of the modern field effect transistors is $0.25 - 2\mu m$. However, in the laboratories the devices are under trial,

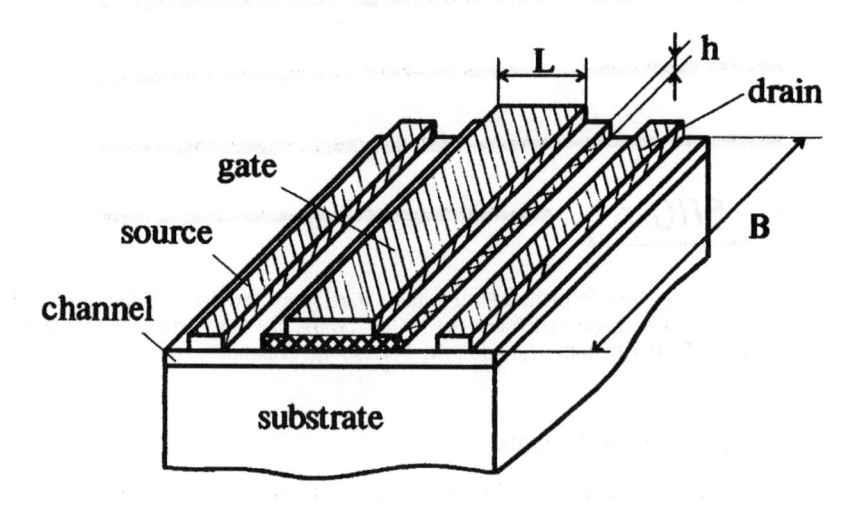

Fig. 92. Typical design of MOSFET.

whose gate length is much smaller, about $\approx 0.05 - 0.1 \mu$m. The width of the gate B is usually $10 - 200 \mu$m. Thus, the input capacity of the gate is within the range of $0.05 - 1$ pF.

The input impedance of the MOSFET is equal to $X_c = 1/\omega C_g$. With the direct current input resistance R at about 10^7 Ohm and $C_g = 1$ pF, the the capacitive impedance will become smaller than the active resistance at the frequency $f \geq 20$ kHz. With any further growth of frequency, the MOSFETs input impedance will be determined exclusively by the capacitance of the gate.

8.5 The Work of the FETs in Actual Regimes

So far, discussing the principle of work of the FET, we in fact thought of the transistor as a resistor whose value can be changed by means of the bias applied to the gate.

Let us take a JFET with a p-n junction as an example.

When no voltage is applied to the gate ($U_g = 0$), the thickness of the channel is maximal. The resistance of the channel $R_0 = \rho L/S = \rho L/Bd_0$. Here $\rho = 1/\sigma = 1/q\mu n_0$ is the resistivity of the material of the channel, d_0 is the total thickness of the channel. If the voltage U_{ds} is applied between the source and the drain, there will be the drain current $I_d = U_{ds}/R_0$.

If the reverse bias $U_g = U_1$ is applied to the gate, the thickness of the channel will be decreased by the value of the space charge region W_1 (See Fig. 86). Knowing the concentration of electrons in the channel n_0 (it's evidently equal to the concentration of donors N), the value W_1 can be easily calculated [see Eq. (40)]:

$$W_1 \cong \left(\frac{2\varepsilon\varepsilon_0 U_1}{qN}\right)^{1/2}{}^{\text{a}}$$

Now the channel will present the following resistance: $R_1 = \rho L/S_1 = \rho L/B \cdot (d_0 - W_1)$. With the drain-to-source voltage U_{ds} the current will be equal to $I_d = U_{ds}/R_1$.

With the gate voltage $U_g = U_2 > U_1$ the resistance of the channel will increase more. Finally, with the pinch-off voltage being $U_g = U_p$, the thickness of the channel $d = d_0 - W_1$ will be reduced to zero. The resistance of

---

[a]Here and below, we will neglect the width of the depletion region with the gate voltage being equal to zero, and in Eq. (40) we will neglect the value of U_{pn}.

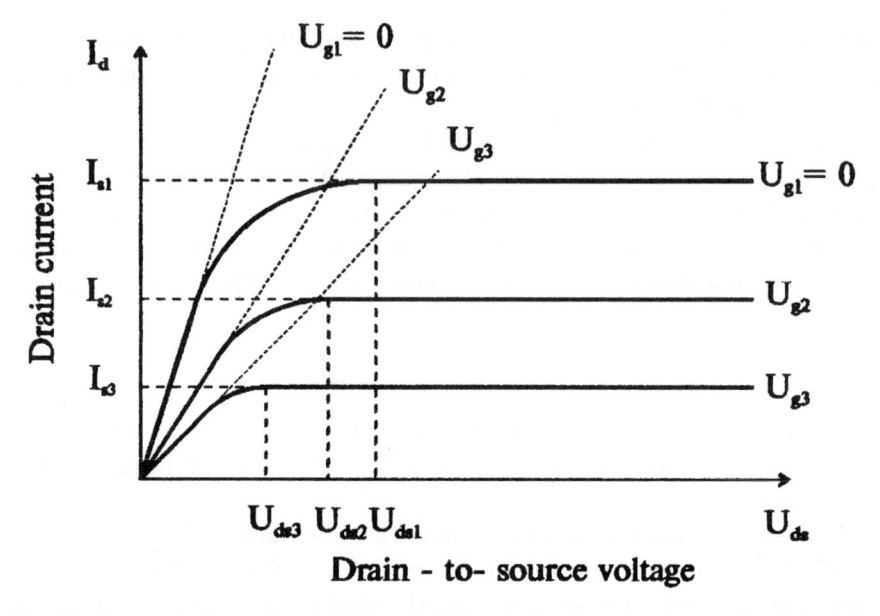

Fig. 93. Current-voltage characteristics of FET at different gate voltages U_g Dashed lines represent the case the transistor's channel is considered simply as a resistor. The value of this resistor is controlled by the gate voltage U_g. Solid curves are the actual current-voltage characteristics of FET.

the channel will become enormous. In Fig. 93, the dashed lines indicate the current-voltage characteristics of the FET, which refer to the picture we have considered.

Solid lines indicate the actual dependences of the drain current I_d on the drain-to-source voltage U_{ds}. One can see that at very small drain-to-source voltage, the actual characteristics coincide with the idealised characteristics. But in case of large currents, i.e. when U_{ds} is large, the actual characteristics have nothing in common with those shown by the dashed lines. The actual current-voltage characteristics *are saturated*. The drain current I_d does not depend on U_{ds} any longer and gets saturated at the I_s level.

The greater the voltage at the gate U_g, the smaller the saturated current I_s and the drain-to-source voltage $U_{ds} = U_s$ at which the value $I = I_s$ is reached (Fig. 93). The FETs of any kind in the overwhelming majority of cases operate in the regime when the current I_d does not depend on the drain-to-source voltage U_{ds}, i.e. in the saturation regime.

Later, we will discuss why it is so. For now, we will try to explain why the actual current-voltage characteristics get saturated.

Figure 94(a) shows the design of a JFET which is already familiar to us (compare this with Fig. 86). A small reverse bias U_g is applied to the gate (with respect to the source). The voltage U_{ds} is applied to the drain (again with regard to the source). The picture is quite familiar.

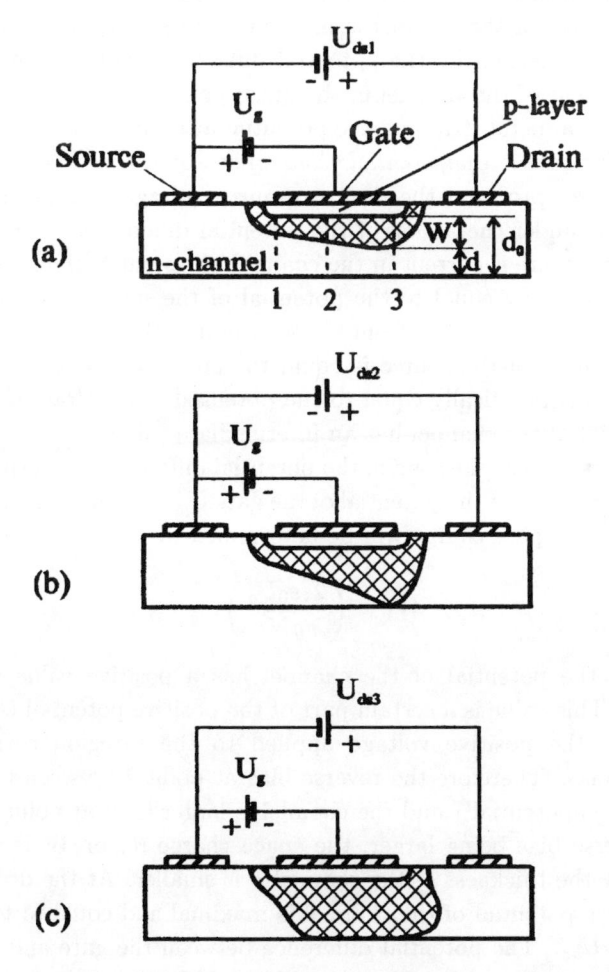

Fig. 94. Current saturation in FET; $U_{ds1} < U_{ds2} < U_{ds3}$ (a) $U_{ds1} < U_s$; (b) $U_{ds2} = U_s$; (c) $U_{ds3} > U_s$; Gate voltage U_g is the same for all three cases.

Earlier we reasoned in this way. The reverse bias U_g forms a space charge region W wide beneath the gate. There is no current in that region. The thickness of the channel is reduced from the value d_0 to the value $d = d_0 - W$. There is current in that channel $I_d = U_{ds}/R$.

But that is not true. Or, to put it mildly, not quite true.

The truth is that when the voltage U_{ds} is applied, the thickness of the channel d is different at different points. At the source the thickness is at its maximum, while at the drain, the width of the space charge region is maximal and the thickness of the channel is minimal. Indeed, the width of the space charge region is defined by the potential difference between the potential of the gate and that of the channel in the given point.

The gate is a metal strip and the potential at each point of it is, naturally, the same. *The gate is equipotential.* And what about the channel?

The channel, provided the drain-to-source voltage is applied, cannot be equipotential: under the action of the potential difference between the drain and the source there is current in the channel. At point 1, the potential of the channel is practically equal to the potential of the source [Fig. 94(a)]. (The voltage being measured down from the source, it is then convenient to assume that the potential of the source is equal to zero). At point 3, the potential of the channel is practically equal to the potential of the drain U_{ds}. At point 2 the potential of the channel has an intermediate value.

So, at point 1, near the source, the potential difference between the channel and the gate is equal to the potential of the gate U_g. It is only at this point that the width of the space charge will be determined by the familiar expression

$$W_1 = \left(\frac{2\varepsilon\varepsilon_0 U_g}{qN}\right)^{1/2}$$

At point 2, the potential of the channel has a positive value with regard to point 1. (This value is a certain part of the positive potential U_{ds}). For the p-n junction the positive voltage applied to the n-region corresponds to the reverse bias. Therefore the reverse bias at point 2 between the gate (the gate being equipotential!) and the channel is larger than at point 1.

The reverse bias being larger, the space charge region W is wider. And consequently the thickness of the channel d is smaller. At the drain, at point 3, the positive potential of the channel is maximal and equal to the potential of the drain U_{ds}. The potential difference between the gate and the channel is also maximal and equal to $U_g + U_{ds}$. The width of the space charge at this point, nearest to the drain, is maximal and equal to

$$W_3 = W_{\max} = \left[\frac{2\varepsilon\varepsilon_0(U_g + U_{ds})}{qN}\right]^{1/2} \qquad (64)$$

So, when U_{ds} is applied, the thickness of the channel becomes different along the channel. It is maximal at the source and minimal at the drain.

The greater U_{ds} becomes, the greater is the "distortion" of the channel. Indeed, the width of the space charge region near the source (at point 1) does not depend on the value U_{ds}. It is determined only by the voltage at the gate U_g. And near the drain (at point 3) the thickness of the channel will get always smaller while U_{ds} is growing.

With the growth of the voltage U_{ds}, the resistance of the channel R is also growing. Indeed, the larger U_{ds} is, the narrower becomes the channel at every point (except point 1 at the very source). And the narrower the channel, the larger the resistance. That is why with the growth of U_{ds}, the current-voltage characteristics (Fig. 93) begin to diverge from the straight line and display a tendency to saturation.

With $U_{ds} = U_s$, the current-voltage characteristics get saturated (Fig. 93). As that takes place the space charge region near the drain practically cuts the channel off [Fig. 94(b)]. Any further growth of U_{ds} does not affect the drain current (Fig. 93) and the length of the cut off region increases [Fig. 94(c)].

The U_{ds} at which the current will be saturated and the channel at the drain will prove to be cut off, depends on the gate voltage U_g. When U_{ds} is small and the width of the space charge region W is practically the same along the channel, the thickness of the channel is defined solely by the value U_g. The larger the U_g, the narrower the channel. And the narrower the channel, the smaller the U_{ds} at which the space charge region will reach the bottom of the channel [Eq. (64), Fig. 94(b)], and the lower the saturation current I_s (Fig. 93).

8.5.1 *The main parameters of FETs*

One of the most important parameters of any transistor is the gain coefficient.

In a bipolar transistor, this parameter is the current gain β (Chapter 7, p. 175). In a FET the gain coefficient of the device is defined by *transconductance*.

Transconductance. The ratio of the drain current change ΔI_d to the gate voltage change ΔU_g is called *transconductance*.

To designate that parameter, the symbol S is usually used

$$S = \frac{\Delta I_d}{\Delta U_g} \qquad (65)$$

Transconductance is usually measured in Ohm^{-1} or in S (Siemens). It is named after the famous German inventor William Siemens.

The greater the transconductance, the greater is the drain current change as a "response" of the FET to the gate signal. As a result of it, the gain coefficient of the FET is greater. Therefore the value S should be made as large as possible.

Look at Fig. 95. It shows the familiar current-voltage characteristics of the field effect transistor (compare this with Fig. 93). Let the gate voltage be zero, $U_g = 0$. Then the upper curve corresponds to the current-voltage characteristics.

Let us first assume that the drain voltage U_{ds} is not large and the current I_d is quite small. Point 1 on the upper curve corresponds to that regime. Let us change the gate voltage from $U_g = 0$ to $U_g = U_2$ without changing the drain voltage U_{ds}. The current I_d will be decreased by ΔI_1. The transconductance in this regime will be $S_1 = \Delta I_1/U_2$.

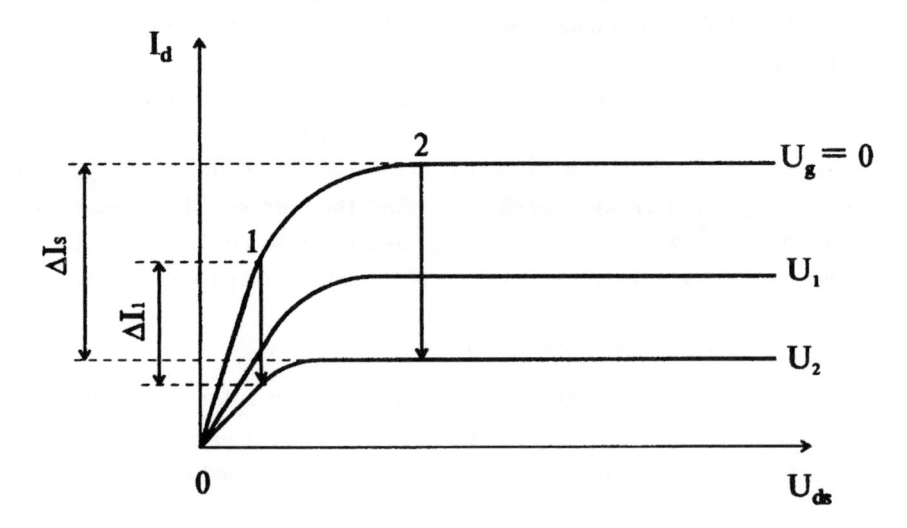

Fig. 95. Current-voltage characteristic of FET (compare to Fig. 93). The same voltage change ΔU_g at different values of U_{ds} provides the different values of the drain current change ΔI_d. In the saturation regime, ΔI_d is maximal.

Let the gate voltage be again zero, $U_g = 0$, but let the drain voltage be rather large, so that the transistor is operating in the saturation regime. Point 2 in the upper current-voltage characteristic corresponds to that regime.

Let us again change the gate voltage from $U_g = 0$ to $U_g = U_2$. The current across the transistor will decrease, but now it will be decreased by the value $\Delta I_s > \Delta I_1$. The transconductance in this regime is $S = \Delta I_s/U_2 > S_1$.

Let us pay attention to the following very important circumstance. As it is clear from Fig. 95, the transconductance of the FET in the saturation regime is at maximum. That is why it was already mentioned in the overwhelming majority of cases where the FET is used in the saturation regime.

The transconductance of the transistor S depends on its design and on the regime of operation: the gate voltage U_g and the drain voltage U_{ds}. In the general case, the equation describing the dependence of the transconductance on all the parameters looks quite terrifying. However, it is possible and quite simple to estimate the transconductance value S in the saturation regime, which is most important.

Let us imagine again that $U_g = 0$. Let U_{ds} be also equal to zero. Then the thickness of the channel in every point will be the same, it will be maximal and equal to d_0. The resistance of the channel is $R_0 = \rho L/Bd_0$. We will leave the gate voltage $U_g = 0$ and will increase the drain voltage U_{ds}. The channel will be distorted — getting narrower at the drain. The current with the increase of U_{ds} will grow, but always more slowly. With $U_{ds} = U_{s1}$ the current will reach its maximum value I_{s1} (Fig. 93, the upper curve). Then, as we know, the space charge will cut off the channel at the drain end [Fig. 94(b)].

One can prove that in such a regime ($U_g = 0, U_{ds} = U_{s1}$) the resistance of the channel will increase with regard to the initial value approximately by a factor of 3: $R_{s1} \approx 3R_0$. Then it is evident that the saturation current I_{s1} will be equal to $I_{s1} \approx U_{s1}/3R_0$.

Let us now apply the pinch-off voltage $U_g = U_p$ to the gate. Then the current flowing across the transistor will evidently fall to zero. And now let us prove that $U_p = U_{s1}$ no matter how unexpected that may seem on the face of it.

It is very simple to prove it. Look at Eq. (64). With $U_g = 0$, the value of U_{ds}, necessary to cut off the channel, must be

$$U_{ds} = U_{s1} = \frac{qNW_{\max}^2}{2\varepsilon\varepsilon_0} = \frac{qNd_0^2}{2\varepsilon\varepsilon_0}$$

On the other hand, the pinch-off voltage U_p is by definition a voltage at the gate which cuts off the channel at $U_{ds} = 0$:

$$U_p = \frac{qNd_0^2}{2\varepsilon\varepsilon_0}$$

(Compare this with the expression on p. 213)
Thus, $U_p = U_{s1}$.

The transconductance of the transistor S corresponding to the change of U_g from $U_g = 0 (I = I_{s1})$ to $U_g = U_p (I = 0)$ is, evidently, equal to

$$S = \frac{\Delta I}{\Delta U_g} = \frac{I_{s1}}{U_p} = \frac{I_{s1}}{U_{s1}} \cong \frac{1}{3R_0}$$

or

$$S \cong \frac{1}{3R_0} = \frac{Bd_0}{3\rho L} = \frac{q\mu N B d_0}{3L} \tag{66}$$

Equation (66) describes quite correctly the main dependences of transconductance on the transistor parameters. The shorter the length of the gate L and the higher the electron mobility μ, the greater is the transconductance of the FET.

The first FETs were manufactured from the n-type silicon whose mobility of electrons is about $0.1 \text{ m}^2/(\text{V·s})$ (See Table 3), while the gate length constituted of dozens of micrometres. The typical value of the transconductance of the transistors made $\approx 10^{-4} S$.

Achievements of modern photolithography made it possible to manufacture the FETs with gate length of decimal fractions of a micrometer. They are made not only of silicon but also of gallium arsenide, whose electron mobility is $0.5 - 0.6 \text{ m}^2/(\text{V·s})$. The transconductance of modern field effect transistors is $\approx 10^{-2} S$. The values of transconductance of the FETs which are being elaborated at the laboratories reach $\approx 1 - 2\ S$.

Speed of response. Analysing the operation of a bipolar transistor, we made sure that it is convenient to characterise the transistors speed of response by the rise time t_0. After changing the input signal, the new value of the output signal is established after the time t_0 (p. 176). The minimal possible rise time of bipolar transistor is defined by the time necessary for the carriers to pass from the emitter across the base to the collector.

Analogically, the minimal possible rise time t_0 in the FET is determined by the time it will take the carriers to pass from the source to the drain. Thus

$t_0 = L/v$, where L is the gate length, v is the mean carrier velocity along the channel. The velocity of the carriers is determined by the field in the channel F. There are FETs in which field in the channel is so strong, that the maximal possible velocity of the carriers is $v_s \approx 10^7$ cm/s is achieved (Fig. 20). For such devices it is, evidently, $t_0 \approx L/v_s$.

With $L \approx 10^{-5}$ cm and $v_s \approx 10^7$ cm/s, the value $t_0 \approx 10^{-12}$ s (≈ 1 ps).

Modern gallium arsenide FETs are semiconductor devices of a great speed of response. The GaAs FETs are capable of generating frequencies over 100 GHz.

However, the FETs in which the field in the channel is much weaker are also used quite widely. The velocity of carriers in those transistors is rather far from saturation and is proportional to the field: $v = \mu F$. The speed of response in such transistors is much less. On the other hand they are very low-power devices. The latter circumstance is especially important in modern computer processors, where the number of the transistors may reach tens of millions.

8.6 FET as an Element of Electronic Circuits

Just like the BTs, the FETs are used in circuits as elements of amplifiers, generators and switchers. The circuit symbols of the FETs (Fig. 96) make it possible to define the type of the FET which is being used.

Details of the circuit design, based on the field-effect and bipolar transistors, are somewhat different, but the main ideas and the general principles are the same. Having studied attentively the basis of the circuit with a bipolar

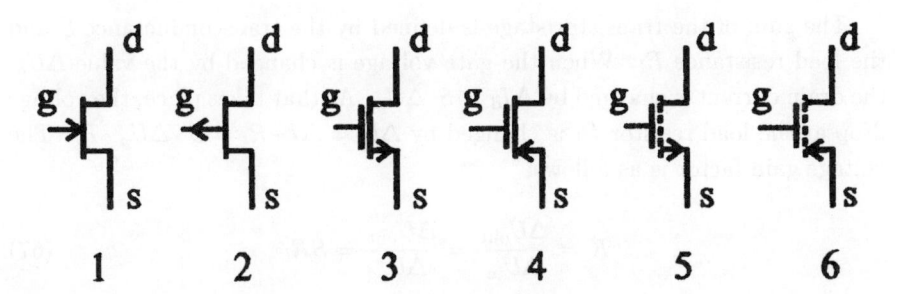

Fig. 96. Circuit symbols of different FET types. s — source; d — drain; g — gate. 1,2 — JFETs; 1 — n-channel; 2 — p-channel; 3,4 — MOSFETS with a normally-on channel; 3 — n-channel, 4 — p-channel; 5,6 — MOSFETS with a normally-off channel; 5 — n-channel; 6 — p-channel.

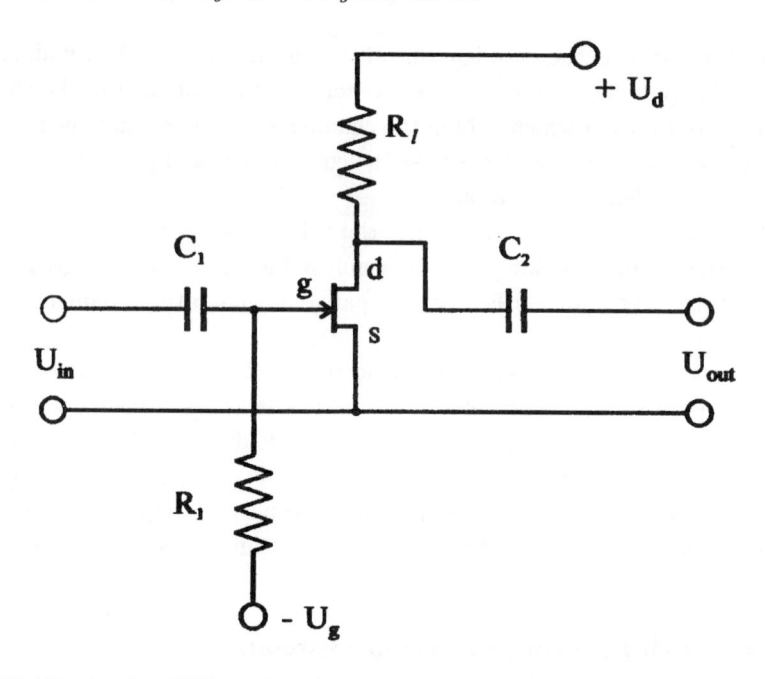

Fig. 97. The simplest FET amplifier (common source configuration) Compare to Fig. 82.

transistor (Section 7.3), one can easily understand the operation of circuits based on FETs.

The load resistance R_l (Fig. 97) is connected in series with the FET, just like it is in the circuit with a bipolar transistor (Fig. 82).

The gain of the transistor stage is defined by the transconductance S and the load resistance R_l. When the gate voltage is changed by the value ΔU_g, the drain current is changed by $\Delta I_d = S \cdot \Delta U_g$. As that takes place, the voltage drop at the load resistor R_l is changed by $\Delta U = \Delta I \cdot R_l = S \cdot \Delta U_g \cdot R_l$. The voltage gain factor is as follows:

$$K = \frac{\Delta U_{\text{out}}}{\Delta U_{\text{in}}} = \frac{\Delta U_{\text{out}}}{\Delta U_g} = S R_l \qquad (67)$$

The value of the resistance R_l is chosen from the same considerations as in the bipolar transistor circuit. As a rule, its value is within the limits from $\approx 10^3$ to $\approx 10^5$ Ohm. With $S \approx 10^{-2} S$, the stage voltage gain is 10 - 10^3.

8.7 Summary

A very simple device makes it possible to amplify electric signals. The device consists of a thin semiconductor plate with two ohmic contacts at the ends, the source and drain, and the third electrode, the gate, placed in the middle and isolated from the semiconductor plate by a dielectric layer.

The voltage, applied between the gate and the semiconductor plate, changes the concentration of carriers in the semiconductor and the resistance of the semiconductor. As a result, the current along the plate changes as well as the voltage drop on the load resistance connected in series with the plate.

The device, *the field effect transistor*, has a very high input resistance and a high speed of response determined by the length of the gate L and the velocity of the carriers v in the channel. The rise time is $t_0 \approx L/v$.

Chapter 9

Transistors and Life

> Pupil: It is much better now
> I see the ways and means
>
> Mephistofel: My friend, the theory is grey,
> But tree of life is ever green and blooming.
>
> *"Faust"* by Goethe J. W. *(1749–1832)*

We began this book with a statement that there are a great many semiconductor devices. We got to know some of them: light-emitting diodes and photodiodes, varicaps and rectifying diodes — each of them is of great practical importance and tens of millions of them are manufactured annually.

And still, none of them can be compared to the transistor either in significance or in the width of applications. For every human being that inhabits the Earth (and they are more than 6 billions) several hundred transistors (on the average) are in operation.

Transistors are said to be the *base elements* of modern electronics.

Though, even before the invention of the transistors, people used to listen to radio, to watch TV, to speak over the telephone to those beyond the ocean, made computer calculations, supplied airplanes with radio-control systems.

So what is the role of the transistor? What has the invention of transistor changed in our life and in the life of humanity?

The brief answer to this question is the following. The transistor can be made of such a small size, so reliable, and consuming so little energy, that on its base it became possible to create such devices and such circuits which cannot be made on the base of any other elements.

A very important and impressive example is the use of transistors in medicine. Tens of thousands of people nowadays use implanted cardio-stimulators. If those stimulators were fabricated not on the basis of the transistors, people would have to carry them in bags or push them on trucks.

It is not necessary any longer to undergo most unpleasant, painful and humiliating procedures. To determine the level of accidity in ones gastric juice, it is enough for the patient suffering from gastritis or stomach ulcer (and there are tens of millions of them) to swallow a small pill and the transistor radio transmitter will transmit the necessary data from the stomach. Every year there appear tens of new applications of electronics in medicine and the decisive role in many of them is played by transistors.

The second part of our brief answer lies in the fact that due to the above-mentioned properties of the transistors — their tiny size, reliability, very small energy consumption — it became possible to improve a lot of devices and circuits! Their being quite tiny, reliable and cheap enormously increased their place and role in our lives.

The next chapter gives a more detailed answer to the questions we have put forth.

9.1 The First King

The electromagnetic waves, predicted by the genius of Englishman James Clerk Maxwell and discovered by gifted German reseacher Heinrich Hertz became a new means of communication due to the works of the wonderful Russian scientist Aleksandr Popov and the splendid Italian engineer and organiser Guglielmo Marconi.

Radio communication which appeared at the end of the XIX century at first was considered to be just a fun source of entertainment. It seemed to be incapable of competing with the wire telegraph communication widely used at that time.

Very soon, however, that attitude to the radio communication changed. And as it is often the case, the first people to understand that radio communication was indispensable and its development inevitable were the military men.

In 1900, the battleship "General-Admiral Apraxin" suffered a shipwreck. It ran aground near a small island of Gogland in the Baltic Sea. It was one of a few Russian ships on which devices of a wireless communication have been installed. They were invented and manufactured by Aleksandr Popov, Professor of the Naval School in Kronshtadt. During the rescue work, the radio-communication was established and maintained at a distance of 30 miles between the battleship and the Finnish town of Kootsali. The availability of that communication played a decisive role in saving the sinking battleship. This accident was broadly illuminated in the press all over the world. It was one of the events which stimulated reseachers to begin working on inventing a device capable of amplifying electric signals and, consequently, increasing the distance of radio-communication.

In 1906, the American investigator Li de Forest applied for a patent on the design of the *vacuum tube* invented by him — the *vacuum triode*. In the next 10 years the design of the triode which had remained principally unchanged was nevertheless greatly improved. During the half a century that followed, the vacuum tube became the base element of the first electronic revolution. At the beginning of the century, however, those dry words were not in fashion. The vacuum tube was then called the King of Electronics.

Amplifiers and generators operating on the base of vacuum tubes were submerged to the bottom of the ocean together with the transatlantic telephone cable and they were flown into the skies as part of the planes radiotransmitters. In 1927, the first industrial television transmitter was created. In the middle of the 1930s, millions of viewers watched the Olyimpic Games taking place in Berlin on their TV. Or they watched the reports from the International Exhibition in New York.

Vacuum tubes provided the progress of the new science *radioelectronics* for many years ahead. It was already quite essential for the physicists, for the medical and military men, for sailors and criminologists and for people of other professions.

The great demand for the vacuum tubes resulted in a lot of improvements in their parameters. Thousands of scientists and engineers throughout the world in dozens of laboratories and institutes worked at perfecting the vacuum tube. Special highly productive automatic lines were designed to produce the vacuum tubes, specialised plants were built to manufacture vacuum tubes. The tubes were being smaller, still more economic and reliable, more efficient and much cheaper.

But years passed and radio-electronics was faced with more complicated problems.

Before usage, the tubes must be heated and that would take several minutes. But those minutes might be crucial for the fate of a plane that had lost control or a battleship whose radar had failed.

The lifetime of the tube is principally limited. The incandescent cathode of the tube, emitting electrons, gradually becomes damaged — and the lamp fails to operate.

When the electron circuit contained just a few tubes, and the lifetime of each tube was rather long, there were no special problems with them. But by the end of the 2nd World War, the radar complexes comprised thousands of tubes. And the computer, which had been designed and had just begun to be manufactured, contained tens of thousands of tubes. The consequences of just one tube failing, the device containing thousands and tens of thousands of components, might be somewhat grave. The same refers to cases when it is very hard to substitute a tube, i.e. in the amplifier of the intercontinental telephone cable at the bottom of the ocean. Consequences may also be very grave.

Devices containing tens of thousands of tubes require a large power station to be fed. Those who work as the electronic systems rack their brains over the problem of how to create new, more complicated devices with the minimal number of vacuum tubes; say, to make the same tube perform the functions of an amplifier, detector, and the generator of signals.

The more complicated the requirements, the more intricate improvements were introduced by the scientists, designers and technologists elaborating the vacuum tubes. The king showed no desire to relinquish the throne.

Ideas were put forward to create a tube "with a cold cathode" which would not need heating. A series of super-reliable tubes was manufactured whose lifetime was guaranteed to be not less than 200 000 hours. Designs of very tiny and super-economic tubes were proposed.

And meanwhile ...

9.2 Ugly Duckling

>...How ugly he is!
>We won't have him!
>
>*The fairy tale " The Ugly Duckling"*
>*by Hans Christian Andersen.*

And meanwhile separate groups of scientists went on working, though the investigations seemed to be fruitless. Again and again, they attempted to create a solid state amplifier of electric signals.

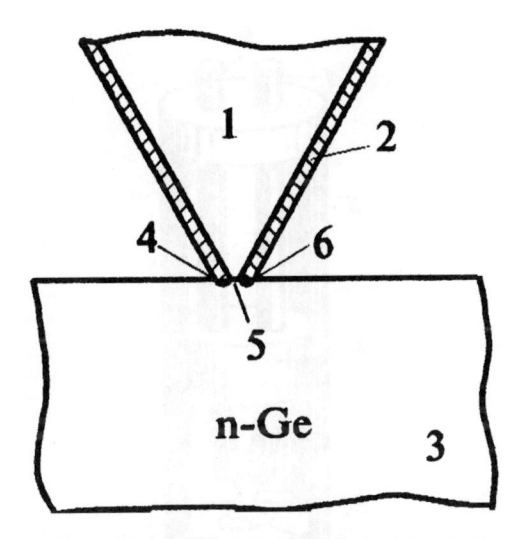

Fig. 98. Design of the first bipolar transistor. The plastic triangle 1 wrapped in gold foil 2 is pressed with force into a piece of the Ge monocrystal of n-type 3. On the triangle vertex, pressed in the plate, the foil is slit with a razor at the place where the foil is contiguous to the monocrystal surface, the p-type regions are formed: 4 — emitter, 6 — collector. Base 5 is located between the emitter and the collector.

Then, in 1947, one group had a breakthrough. W. Shockley, J. Bardeen, and W. Brattain had invented a bipolar transistor.

But dear me! How ugly, puny and frail it was!

The design of the first bipolar transistor is shown in Fig. 98. What "reliability", "reproducibility", "tolerance for outer effects" or "durability" could be expected of such a "home-made" design?

A year later, in 1948, the first samples of the industrial bipolar transistors were on sale (Fig. 99). Alas, their reliability, reproducibility, etc. did not differ much from the experimental design shown in Fig. 98. It was enough just to shake such a "device" and its voltage gain would change a lot (by a factor of several times).

When manufacturing such a transistor, one could never predict what its gain might be. Should the temperature be slightly raised or lowered, the transistor would stop working.

That ugly duckling could not compete with the King for whom thousands of servants had been working, the industry of vacuum tubes having been developed and improved for 30 years.

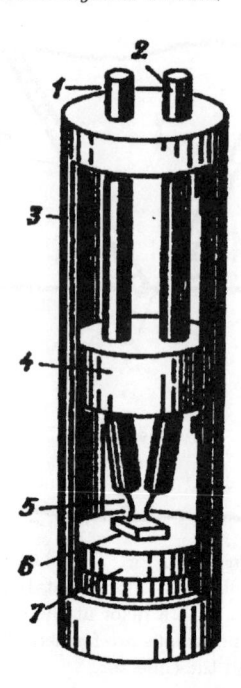

230   Transistors. From Crystals to Integrated Circuits

Fig. 99. The first industrial BT (1948). Emitter and collector p-n junctions are fabricated by alloying of thin wires into Ge crystal. The diameter of the transistor package is 1 cm; the height of the package is 4 cm. 1 — emitter lead; 2 — collector lead; 3 — package; 4 — insulator; 5 — two contact wires; 6 — Ge crystal; 7 — base lead.

But the "ugly duckling" did his best. He tried as hard as he could. Almost simultaneously with the appearance of the first industrial samples of transistors, experimental specimens of transistor radioreceivers, TVs and even amplifiers for the guitar were created and demonstrated.

Alas, the poultry yard was not affected by all that. When the firm "Bell Laboratories", where the inventors of the transistor were working, inquired the military men whether the new device should be classified as a secret. The military turkey-cocks, having just cast a look at the new device, then allowed any kind of publications about it.

Neither public demonstation of the device, nor advertising attracted any attention to it. Nevertheless, small experimental batches of the transistors were being produced.

The first consumers appeared very soon. Those were the firms producing the apparatus for the people with reduced hearing. The greatest advantage of the transistor was the tiny size of transistor circuits. They need a very low battery voltage (2–3 V), and the transistor amplifier placed in the rim of glasses was absolutely safe. The Bell Company sold such transistors at a very low price — in honour of Alexander Bell, who taught at the school for the deaf and dumb.

The second group of consumers was the military personnel. At the very end of the 1940s, they began to develop on the systems capable of launching the man made satellites of the Earth. According to the calculation, every gram of the load launched into the orbit was to cost so much, that it made sense to work at improving those capricious, but much lighter, small and economic devices. The military departments did not suffer from lack of funds and acted on a wide scale. They bought large batches of transistors by wholesale and put them on various trials, comprehensive and rigid. They were shaken on vibration-testing machines, heated in thermostats, submerged into the water and tested with increased loads. Out of 50–100 transistors, one was selected which had gone through fire and water, and based on this transistor, circuits were designed.

Meanwhile the parameters of the transistors were rapidly improving. Alloyed transistors have evolved, reliable, with reproducible parameters, resistant to outer effects. Methods of serial production of transistors have been elaborated and they became much cheaper. Now the military firms as well as the civil firms were willing to buy them — the transistors being cheaper and more reliable.

Transistors became profitable. Part of the profit could be spent on developing new technological processes, on creating the new highly-productive lines and developing new types of transistors. The demand for those improved and cheap transistors was always growing, the circle of consumers widening.

9.3 Long Live the New King!

Only six years after the transistor has been invented in 1953, one of the most popular technical journals in the world, "Electronics", wrote that since humanity was entering a new era, the era of transistors, it was necessary that the engineers would make a deeper analysis of the potential of those new devices so similar to vacuum tubes and yet so different. Transistor circuits can be improved upon even if the transistor is assumed to be a device substituting a tube. But in order to achieve new results, it is essential that the fundamental peculiarities of the transistors should be used.

The period of "getting accustomed" to the transistors lasted the first half of the 1950s. At the beginning of that period, the transistor circuits, replacing the similar tube circuits, were being improved upon. By the beginning of the 1960s, engineers at last began to appreciate the main features of transistors: small power consumption, a tiny mass, very small sizes, and a very high reliability. In 1954, about 5 million transistors were produced, in 1958, about 200 million, in 1963, about 1.5 billion! In 1963, about 2500 types of diodes and 3000 types of transistors were produced in the world.

The old King — the vacuum tube — after reigning for half a century renounced the throne.[a]

Transistor amplifiers and generators made base elements of the new generation of TVs, radios and tape recorders. Combined with other semiconductor elements, the transistors serve as exact thermometers and electronic scales. High-voltage transistors control powerful electric engines. Special transistors have been developed to control the ignition system of the cars. Supersensitive, super high frequency transistors receive the signals from the artificial satellites of the Earth and from interplanetary spaceships. Portable transistor apparatus transmit over the telephone the patients cardiogram directly from the patients home to the doctors office. Tiny sensors control the patients pulse and blood pressure and give information about the internal organs of a person.

During the 10 years, at the beginning of the 1960s, the range of application of transistors widened to such an extent that a mere enumeration of their functions would take several thick volumes.

One of the most attractive possibilities afforded to electronics, due to a high reliability and efficiency of the transistor, is to create very complicated circuits, containing tens or even hundreds of thousands of elements. It was the transistors that became the base elements of the second generation of computers.

Such an abrupt complication of circuits raised a new problem, one that had never been dealt with before. The reliability of a vacuum tube was so much lower than that of any other elements that only the number and the reliability of tubes determined the general durability and reliability of the electronic equipment. That is quite different with a transistor!

---

[a] However, even nowadays hundreds of millions of vacuum tubes are being produced. The tubes remain indispensable where large power and super high frequencies are wanted.

It is not less durable than any other element of the circuit: capacitor, resistor, a piece of wire, the contact between the elements, etc. Just like 40 years ago, life raised new problems before electronics. It was necessary to produce still more complicated circuits. But no improvements of the transistor's parameters could help any longer. It took only one piece of wire to be torn, or one resistor to fail (it was next to impossible to find and spot them in the circuit), and the system would not work.

9.4 The King ... Disappears. Long Live the New King!

In 1952, the annual conference on the elements of electron circuits was held in Washington, DC. It goes without saying that the majority of presentations were devoted to tube circuits. But one of the participants, Jeffri Dammers, an engineer from Great Britain, made a report which had nothing to do with the actual problems of electronics. He remarked that *with the appearance of a transistor and semiconductor technology, the electronic equipment can be assumed to be a single solid state block without any connecting wires.* The block may consist of insulating, conducting, rectifying and amplifying layers.

At that time, Dammer's voice sounded singular in the wilderness. Few people were interested in transistor circuits. And the problem of their reliability was to no interest to anyone.

But 10 years later, when it was taking months for thousands of qualified specialists to assemble a computer, and when just one unreliable contact might cause the whole most complicated apparatus to fail ...

Do you think they recalled Dammer's words? Nothing of the kind! In 1952, Dammer's presentation did not attract anyone's attention and was swiftly forgotten. It was only a quarter of a century later that historians of science came across it. More and more people came to the same conclusion which Dammer so miraculously predicted.

By that time most of the transistors were produced by methods of diffusion and alloying, according to the planar technology. In that case all the transistors on a semiconductor plate were manufactured simultaneously. Why then should one, after the transistor structures were made, cut that plate into separate elements, mounting them into packages, bounding the wire leads and make a lot of other complex and tiring operations, obtaining the ready-made transistors ... and again, applying enormous efforts, to combine the transistors into a complicated circuit? Isn't it more natural and expedient to connect all the transistors on the semiconductor plate, thus obtaining a ready-made circuit?

An integrated circuit!

The first intergrated circuit (IC) was on sale in 1961. It presented a transistor trigger and contained 4 BTs and 2 resistors — 6 elements in all, located on a silicon plate 1 cm in diameter.

Two years later, in 1963, the number of elements in the IC increased to 10–20, the size of it decreasing to 3–5 mm. By 1967, the minimal number of the elements on the crystal made ~ 100; by 1970, ~ 1000; by 1975, $\sim 30\,000$; by 1982, 300 000 and by 1996, 10 000 000!

As a matter of fact, we are already acquainted with the principle of manufacturing the intergrated circuits. It does not differ at all from that, already described by us, principle of a simultaneous manufacturing (group manufacturing) on the silicon wafer of several dozen transistors or even hundreds of thousands of transistors by methods of planar technology. To join transistors to each other, evaporated or chemically deposited strips of metal are used. Aluminium is often used for this purpose. The deposited strips of material with the assigned resistivity, length and thickness are used as resistors. The reverse biased p-n junctions are used as capacitors in ICs. As for inductivity, they try to avoid them in ICs. When greatly magnified, the surface of the silicon plate with IC manufactured on it, looks as it is shown in Fig. 100. A complicated geometrical ornament is formed on the surface of the semiconductor wafer by the transistors, metallic strips, and the insulating layers of the oxide.

The *level of integration*, i.e. the density of elements of IC on the semiconductor wafer is such that the elements cannot be bonded by hand, no matter how skilled those hands might be.

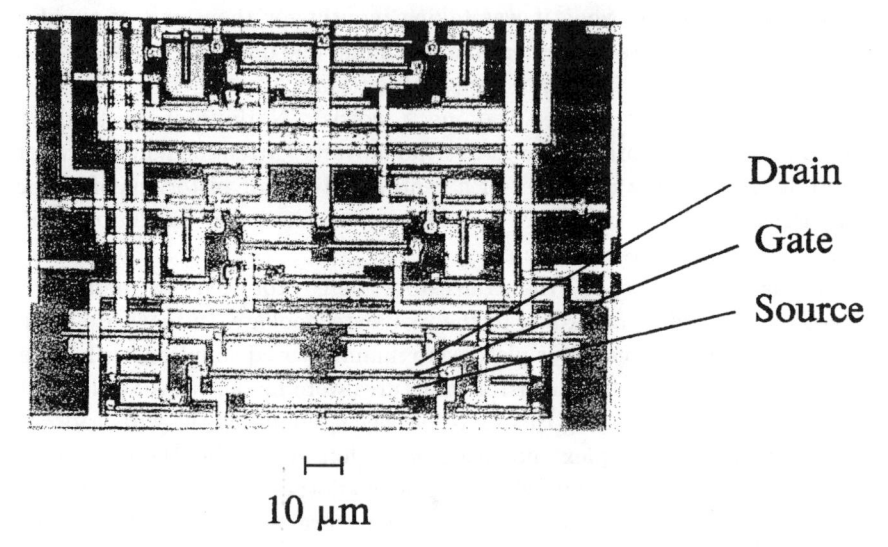

Fig. 100. A small part of a modern high-frequency GaAs IC based on FETs. The gate length of FETs is about 1 μm.

The whole cycle of operations of creating IC goes on in closed chambers where not a speck of dust can penetrate. Nothing is made by hand. Human hands do not take part in that process. The modern technology of producing IC, LSI and VLSI (Large Scale and Very Large Scale Intergrated circuits) besides small sizes provides exceptionally high reliability of complicated radio-electronic systems.

The first IC, containing several dozen elements, were indicated in the circuits just like ordinary circuits on the base of discrete elements. The only difference was that the elements which belonged to IC were enclosed in a dashed frame [Fig. 101(a)]. However, the more intricate were ICs, the more elements they contained, the simpler were their circuit symbols [Fig. 101(b)].

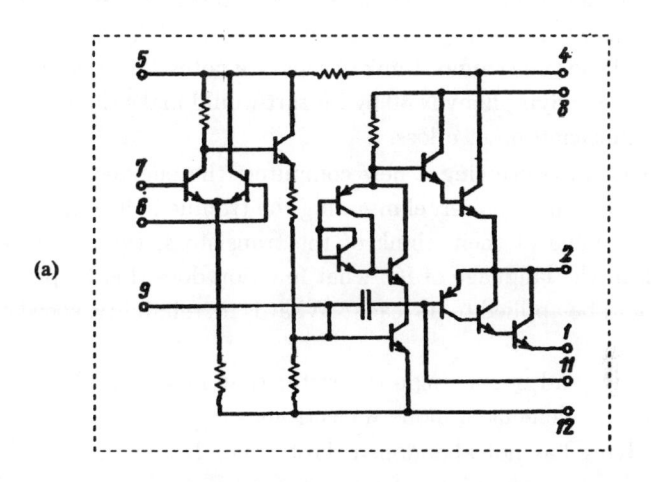

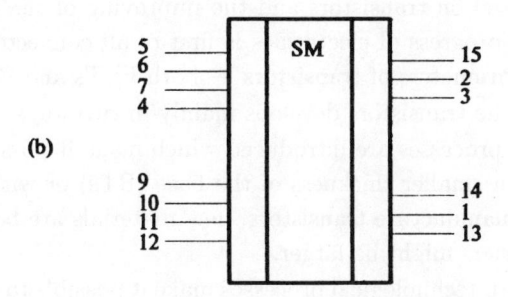

Fig. 101. Schematics of ICs. (a) Schematic of the simple IC containing only several dozen elements. (b) Schematic of the complicated IC which can contain up to ten million elements.

These changes in the designations reflected deep and complicated changes in the thinking of engineers, scientists, producers — everyone who dealt with the ICs.

A single transistor was no longer a base element of electronics! It had been replaced by IC.

It may seem that this statement is a kind of propaganda without any deep idea. Indeed, IC is based on the same FETs or BTs. So is it worthwhile forcing open an open door?

Yes, in this case it is. The designer constructing a modern machine has no idea how many screws or nuts, gaskets or springs it has. Thinking about the construction, he conceives whole blocks: carburettor, chassis, engine, etc. He thinks about their reliability, parameters, sizes .... The musician composing or performing the tune cannot think of separate notes. The question: how many notes there are in Symphony N 40 by Mozart would make the most experienced and skilled musician be at a loss.

The scientist, conceiving a new computer; the engineer, designing a new electronic block; the repairer, eliminating the trouble in a complicated modern apparatus — none of them think of the transistors, the components of IC. They think in the language of IC: what function does this IC perform? What voltage should be applied to its leads? What is its reliability, speed of response, etc?

It is the IC and not a single transistor that makes the base element, the main functional element of modern electronics.

So, the King has not abandoned the throne. He, in the best traditions of democracy, became one of its citizens, one of hundreds of thousands of them. And nevertheless, long live the King!

Still, any IC is based on transistors and the improving of the parameters of IC and the further progress of electronics is first of all connected with the improvement of the parameters of transistors — both FETs and BTs.

The improving of the transistors develops mainly in two ways:

New technological processes are introduced which make it possible to produce transistors with a smaller thickness of the base (BTs) or with a shorter channel (FETs). To manufacture transistors, new materials are being sought whose mobility of carriers might be larger.

As we know, modern technological processes make it possible to grow under the computer's control, the semiconductor layers whose thickness is only a monoatomic layer. Devices with the size of active elements of 10–15 Å have been practically created and are under trial. There are BTs and FETs with the width of the base (length of the channel) of 0.05 μm.

New designs of BTs and FETs are being created. High Electron Mobility Transistors — HEMTs — can be cited as an example. Their operation is based on the properties of a junction which appears on the boundary between the two different semiconductors, the so-called *heterojunction*. We did not study the properties of heterojunctions. But the physical idea which makes up the basis of HEMTs is so ingenious that we can't help saying a few words about it.

As we know one of the main, most important problems is the increase of the speed of response and of the limiting operational frequency of transistors. Scientists do their best trying to develop new types of such transistors. In the FETs operating in the regime of unsaturated velocity, the higher the mobility, the higher the limiting frequency. In BTs the limiting frequency f_c increases with the increase of the diffusion coefficient D, i.e. in accordance with the Einstein relation (22), the limiting frequency increases with the growth of mobility.

As we know, the mobility depends on the mean time between the acts of scattering τ_0. The larger the τ_0, the less frequent are the collisions, the higher is the mobility [Eq. (15)].

One of the main collision processes is the impurity collision. The smaller the impurity concentration, the higher the mobility. Thus, it seems as if it were clear that the right way of increasing the mobility is to remove impurities.

However, things are not so simple. To produce the FETs and BTs, it is necessary to use the semiconductor material with a rather high carrier concentration ($\sim 10^{17}$ cm^{-3}). And as we know, the free carriers appear at a predetermined temperature only provided impurities are introduced into the semiconductor. So we have a vicious circle: if you want to have a desired concentration of carriers in the semiconductor, you should introduce impurites, but as soon as you introduce impurities, the mobility will be decreased.

The idea of the structures, serving as a basis for HEMTs, the so-called *modulation-doped structure* consists in dividing in space the impurities and the carriers which come into being due to those impurities. The impurity (say, the donor impurity) is introduced into the layer of material 1 (Fig. 102). The electrons, which appeared due to that impurity, diffuse into layer 2, located in the immediate proximity. No impurity is introduced into layer 2. And this layer 2, containing electrons but not containing any impurity, serves as an active element of the transistor. It may serve, for instance, as FET's channel.

It goes without saying that it is not any two materials that are suitable for making such a "sandwich". To do credit to the reseachers, they understood what two materials could form a match with the desired properties.

Historically, the first pair of the materials matching each other were GaAs and the ternary compund GaAlAs.

FETs with unique properties were created on the basis of modulation-doped structures using the properties of heterojunction GaAs/GaAlAs. Their switching time is ~ 1 ps (10^{-12} s), and the energy consumed in switching is only ~ 1 fJ (10^{-15} J). What enormous gain in mobility the use of these modulation-doped structures allows to get can be seen from Fig. 103. Figure 104 shows a FET, based on the modulation-doped structure. With the concentration of electrons in the channel being $\sim 2 \cdot 10^{17}$ cm^{-3}, their mobility

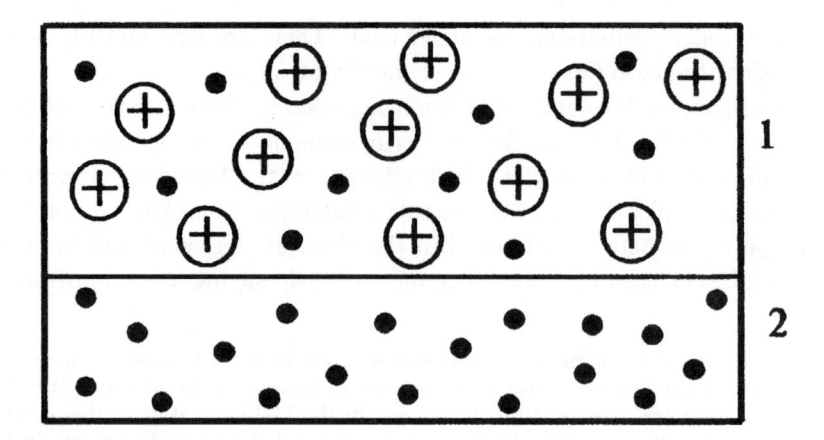

Fig. 102. Modulation doped structure. Donor impurity atoms $\oplus$ are introduced into layer 1 (GaAlAs). These atoms create conduction electrons • in the layer. Some of the conduction electrons diffuse into layer 2 (GaAs). In the GaAs layer with high concentration of high mobility electrons is formed.

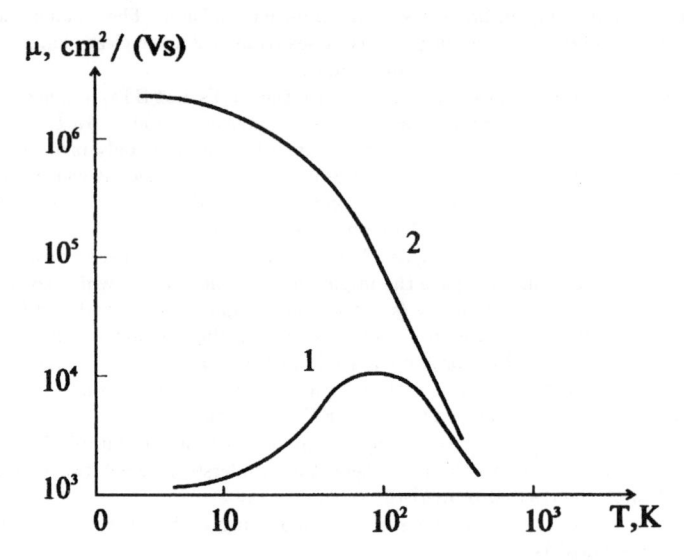

Fig. 103. Temperatures dependence of electron mobility. 1 - for GaAs with the carrier concentration $2 \cdot 10^{17}$ cm^{-3}; 2 - for modulation doped structure with the same concentration.

is approximately twice as large as in the gallium arsenide transistor at room temperature (300 K) and by a factor of ~ 100 larger at a temperature of 80 K.

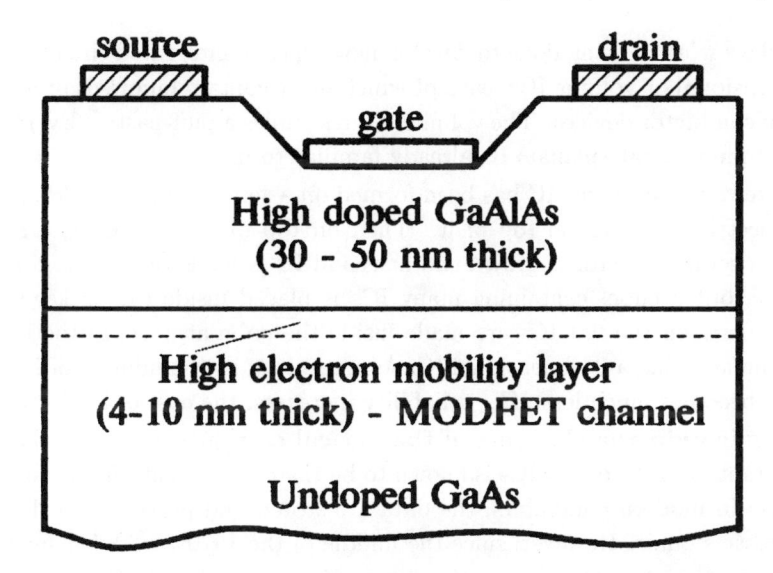

Fig. 104. Modulation doped field-effect transistor (MODFET). At the interface of the highly doped GaAlAs and the low doped GaAs a channel with a high electron mobility is formed.

9.5 Claimants to the Throne

According to the opinion of the majority of specialists in semiconductor physics it will be ICs based on the FETs and BTs that will make the base elements of modern electronics during the coming years.

However, nowadays, just like it was half a century ago, and how it used to be in the history of science, certain groups of researchers try to develop such projects whose realization promises not the improvement of the existing devices, but the creation of new ones. No one can say today which of those ideas may prove to be wrong, and will be forgotten, leaving no followers, which may require scores of years for their implementation, and which in the nearest future may lead to a new revolution in electronics.

In the conclusion of this book, we will tell you about some of those ideas which seem to be most promising today.

Three-dimensional intergrated circuits. Today more than ten million transistors can be successfully placed on a single VLSI silicon chip. But for the electronics of the future, specifically for the future generations of computers even that enormous density of packing will prove to be insufficient. Therefore nowadays work is being done to develop new types of integrated circuits — the 3-dimensional or volume ICs, each of which must contain hundreds of millions of semiconductor devices. The volume ICs resembles a puff-pastry. Every layer of that pie is a flat (planar) IC already familiar to us.

After a usual planar IC has been formed on a semiconductor wafer, a layer of dielectric is grown on top of it. Then on top of the dielectric, a second layer of semiconductor is grown and a second IC is made on it, etc. Now not a plate, but a cube, containing many ICs is placed inside the package. The bonds between the flat ICs are made inside the cube at once, in the process of manufacturing a 3-dimensional IC. As a result of it, the number of external bonds becomes much less. The reliability increases, the cost of IC decreases.

Semiconductor elements of the optical computers. One of the most important applications of ICs is known to be their implementation in the computers. In modern computers, the data are stored and processed in the form of electric signals. However, since the middle of the 1970s, work has begun on creating computers of quite another type — the optical computers.

In modern computers the on and off states of the transistors arbitrarily correspond to "zero" and "unity" — the two numbers by which all the stored and processed information is recorded.

In optical computers which are being designed, "unity" corresponds to the state when light passes through the optic cell, and "zero" corresponds to state when the light is absorbed in the cell. The simplest image, corresponding to "0" and "1" in the optical computers is quite illuminating — it is an electric lamp, switched "on" or "off". A semiconductor laser in the optical computer serves as such a lamp. That laser is the source of light whose intensity can be changed with a frequency of billions of Hertz (Hz). However, the semiconductor lasers consume hundreds of thousands of times more energy than the transistors. Therefore the computer based on hundreds of millions of lasers would consume intolerably much energy.

The beam of a semiconductor laser introduced into the optical guide is supposed to be directed to the semiconductor optical cells, whose transparency can be changed tens of milliards of times per second with a small comsumption of energy.

Bioelectronics. The nature prompts to us that the increase in the speed of response is not the only way to perfect complicated electronic devices, particularly computers. The cell of the human brains — the neurons — switch on and off one million times slower than the transistors — the cells of modern computers. But it was Man who invented the computer, not the other way around.

What is the structure of the human brain? How does it work with such a wonderful efficiency? These questions are being investigated and are still far from being solved. However, the first steps to create biomolecular ICs have already been made. A patent has been obtained to produce IC based on the protein material, polylizin. Molecular switches, polymer FETs, and biosensors of nuclear, biological or chemical processes are made. The density of packing the elements in such biopolymer structures is supposed to approach that in the structures responsible for the photosynthesis in the green leaves of plants and will be over 1 billion per 1 mm^2. There is every reason to expect that one day, the biomolecular ICs will be not manufactured, but grown like a natural tissue. If those hopes come true, such elements will have nothing to do with semiconductors.

As we can see, the vanguards of modern electronics carry out reconnaissance in different directions. But meanwhile it is the progress in the study of the properties of semiconductors and the properties of junctions that will in the next ten years determine the progress in the field of electronics and, to a great extent, in all modern civilization as such.

Conclusion

Step by step, we have covered together the long and complicated road from a crystal to an integrated circuit. And we made sure that "the devil is not so terrible as he is painted." Even such intricate devices as BTs and FETs are based on simple and clear ideas. Those ideas become especially clear and simple if one should think of them, understand them well and put them to practical use.

We have overcome several high and steep "barriers" along the way. The properties of those potential barriers which arise on the interface crystal-vacuum, on the junction between the p- and n-regions of the semiconductor determine to a great extent the work of the most important semiconductor devices. Overcoming these "barriers", we have learned a lot. We truly hope that the skills you acquire when overcoming the barriers, will help you in the future, in the physics of semiconductors or in quite other fields of knowledge. In any case, good luck, and happy journeys ahead!